Arduino LED Cube Projects

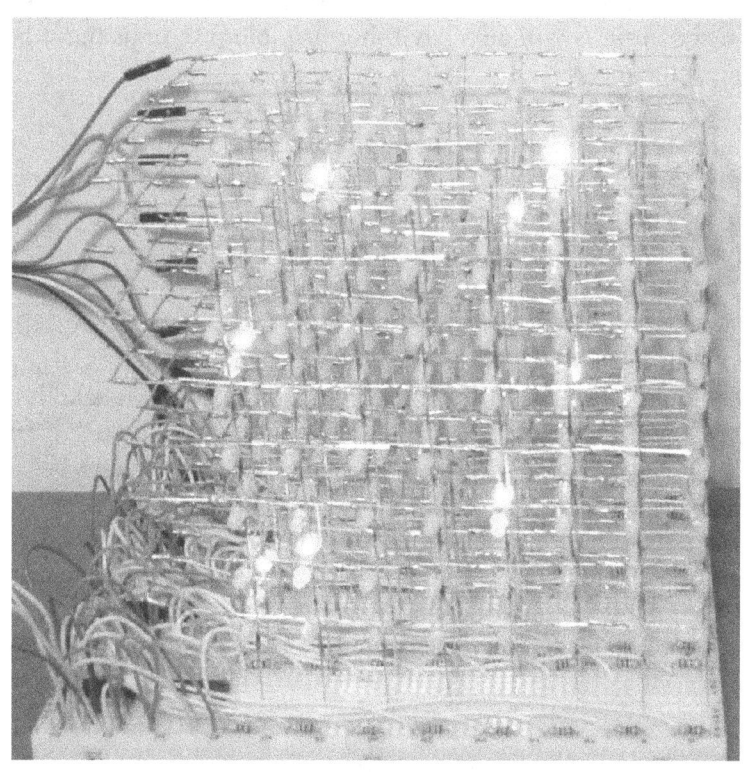

Robert J Davis

Arduino LED Projects 2
By Robert J Davis II
Copyright 2015 by Robert J Davis II

The first book of "Arduino LED Projects" started with individual LED's and ended up with LED signs. Since then I have discovered the fun of making LED cubes. There are so many variations of LED cubes available that there can now be a book written that is just dedicated to how to make the most common varieties of LED cubes.

The LED cubes in this book have been designed to be compatible with and have been tested with the software that is available for the many similar LED cube designs that can be found on Instructables and elsewhere on the Internet. My LED Cube designs will also have short sample programs that I have written to test them out.

As always the reader or builder takes all responsibility for the safe construction and operation of these devices. If you are not familiar with building these complex projects it is highly recommended that you start off with the first book of "Arduino LED Projects".

These projects are fairly complex. I have tried to make them look like they are easy to build. Every design in this book has an explanation, a schematic diagram, a picture of the assembled project, and the code to make it work. This book was written to get you started, by adding more layers of LED's, or customizing the code you can make LED cubes a lot more fun.

Most of the parts to build these projects will need to be ordered from China via eBay or one of the mail order parts vendors like MCM Electronics. Some of the parts are available at your local Radio Shack store. Some of the parts used were just lying around because they were left over from previous projects.

Table of Contents

Chapter 1

Comparing Arduino Versions

There are now many versions of the Arduino. Here are four of the more common versions. They are the Nana, the good old Arduino UNO, the Arduino Mega, and the PcDuino. The Mega added a lot of I/O pins that were needed to interface to some LCD screens and other things that required more than the normal 20 I/O pins. Here is a picture comparing these four processors. The Nano is on the left, then the Uno, the Mega is next and the PCDuino is on the right.

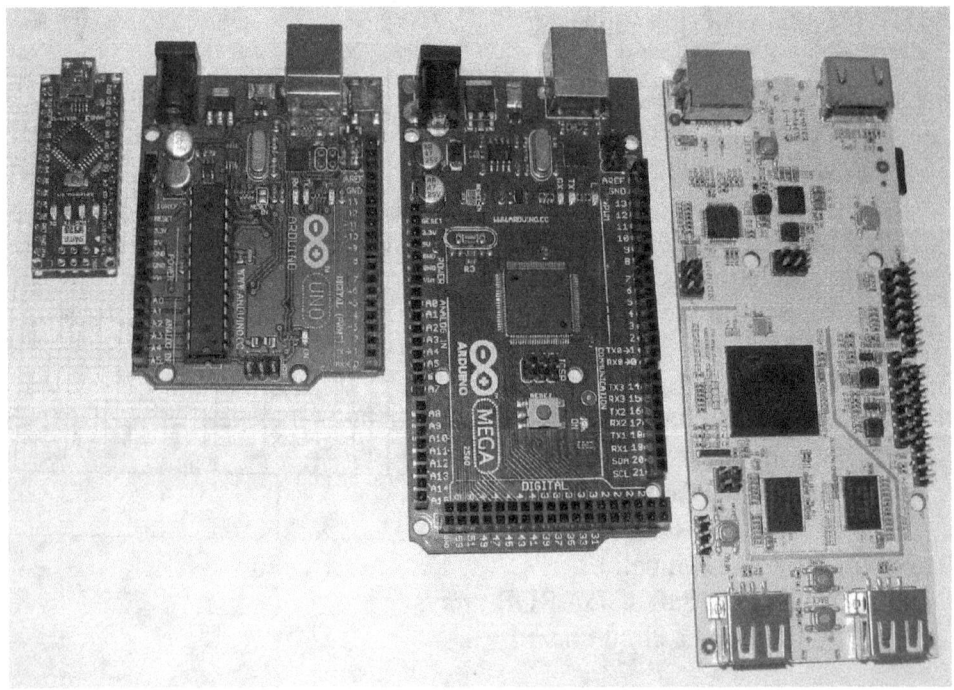

The PcDuino gives you the power of a PC that you can hold in your hand. It eliminates the "middle computer". You can now write the code on the same machine that you run the code.

Here is a comparison chart showing some of the differences between these four versions of the Arduino.

	Nano	Uno	Mega	PcDuino
Clock Frequency	16Mhz	16MHz	16MHz	1GHz
Memory	32K	16/32K	128/256	1Gig+2Gig
I/O Pins	14	14	54	14
Analog Pins	8	6	16	8
Logic Volts	5V	5V	5V	3.3V
Pin Current	40ma	40ma	40ma	4ma
Power Supply	7-12V	7-12V	7-12V	5V
USB Ports?	No	No	No	Yes
Network Jack?	No	No	No	Yes
Video Output?	No	No	No	HDMI
SD Memory?	No	No	No	Up to 32 Gig

The PcDuino can replace the computer that the Arduino applications are normally written on. Within a few minutes needed to familiarize yourself with Linux you can be writing the code on the same computer that you run the code! Once you locate the Arduino icon you can then use the familiar Arduino development environment to run your applications.

Not everything that runs on the Uno can be run on the PcDuino. The most obvious difference is the logic voltage of 3.3 volts and the current limit of 4ma. That compares to a logic level of 5 volts and 40 ma of current for the other Arduino processors.

Here is a picture of a PcDuino mounted on a piece of Plexiglas that is 3 inches by 6 inches in size. It is a good idea to mount it to something otherwise whenever I would handle the PcDuino the SD card was accidentally ejected. Later I added some rubber feet to keep it from sliding around on the desk.

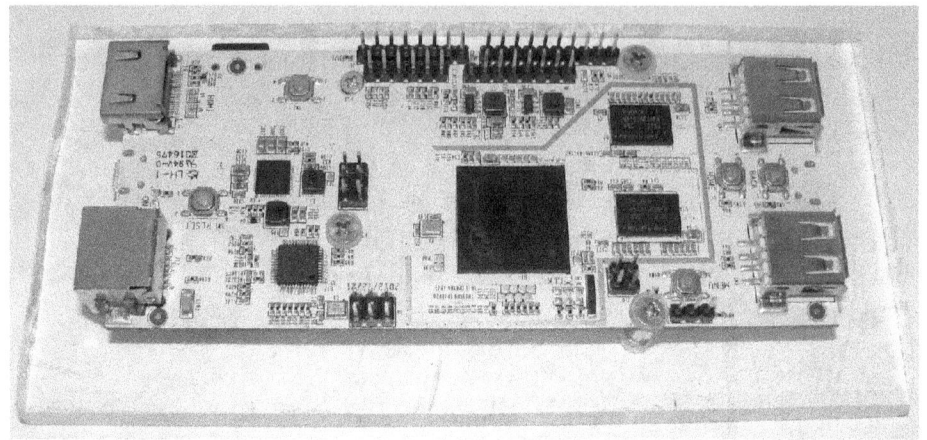

When I first powered up the PcDuino it required that I set the screen resolution. I picked one and the screen went blank and never came back. I re-flashed the SD card and the results were the same. Eventually I found out that the problem was the AC adapter. When they say to use a 2 amp AC adapter they are serious. Apparently there is a power surge when it switches to graphics mode that will have issues with a 1 amp AC adapter.

To use a PcDuino you will need the following accessories:

1. HDMI cable and HDMI monitor.
2. USB Mouse and Keyboard.
3. 5V 2A AC adapter.
4. Optional USB hub to add any USB devices.
5. Optional micro SD card for more room for programs

This next picture shows where to connect things up to the PcDuino.

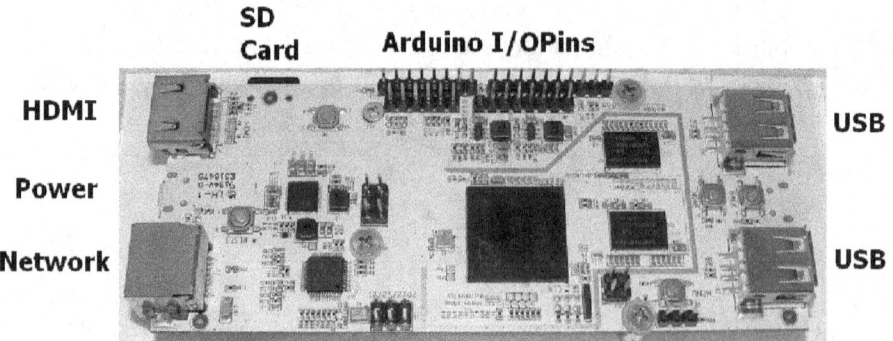

Chapter 2

Introduction to LED Cubes

LED cubes give us the ability to "display" things in 3D. They can be looked at as complex mathematic displays, although they lack a lot of steps as far as showing a lot of detail. You can even display complex mathematical formulas on them.

There are many design considerations to consider. First of all consider the LED. It has leads that are about one inch long. That makes the maximum spacing that uses those leads about one inch. However I tried that and because one lead must go overtop of the other lead it has to be about 1.25 inches long and basically it usually comes up short. You have to either bridge the gap with the solder or use a wire.

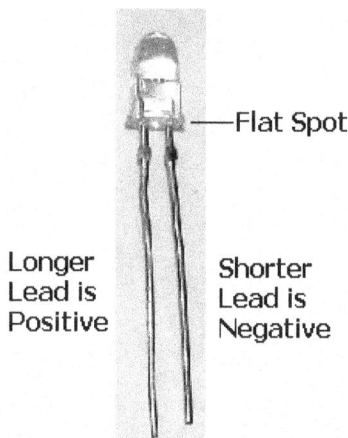

Flat Spot

Longer
Lead is
Positive

Shorter
Lead is
Negative

The easiest LED spacing is .8 inches or 2.0 cm. The 8x8x8 LED cube uses 15cm or .6 inches because it had to be compatible with a circuit board that is part of a kit. However the .6 inch spacing almost forces you to cut one lead shorter as it is too long.

Another design consideration is what you can fit on a breadboard. You need to add a shift register for every 8 LED's as a minimum. You will also need to consider how LED's usually come in packages of 100. Some venders packages are exactly 100 and some will throw in some extra LED's. Here are the mathematics of the common LED cube sizes.

3x3x3=27 LED's
4x4x4=64 LED's
5x5x5=125 LED's
6x6x6=216 LED's
7x7x7=343 LED's
8x8x8=512 LED's

A typical 8x8 LED layer looks like this chart. However if you are using a shift register then the first data bit (left most) gets shifted down to the last shift position (64) so the numbers are reversed in the shifting process.

```
64  63  62  61  60  59  58  57
56  55  54  53  52  51  50  49
48  47  46  45  44  43  42  41
40  39  38  37  36  35  34  33
32  31  30  29  28  27  26  25
24  23  22  21  20  19  18  17
16  15  14  13  12  11  10  9
8   7   6   5   4   3   2   1
```

Mathematically then the X and Y axis would look like this.

```
Y axis
7   o   o   o   o   o   o   o   o
6   o   o   o   o   o   o   o   o
5   o   o   o   o   o   o   o   o
4   o   o   o   o   o   o   o   o
3   o   o   o   o   o   o   o   o
2   o   o   o   o   o   o   o   o
1   o   o   o   o   o   o   o   o
0   o   o   o   o   o   o   o   o
    7   6   5   4   3   2   1   0  X axis
```

It is oriented this way so that the arrays in the software will line up with the actual LED's. That is to say that the first "0" or "1" will end up in LED 64. This makes "reading" the array much easier to do.

The layers of the Z axis generally start with a "Z" of "0" at the top layer and technically going -1, -2, -3, -4, etc as you go down through the layers. However they are usually formulated in software as +1, +2, +3, +4, etc. as you go from the top to the bottom. This might result in some formulas being displayed upside down. In fact I have noticed some shifting of the + and – of the X and Y axis around in various cube designs. If you are just looking for artistic expression then this reversing of an axis is not a serious issue.

So to move right and left in the X axis you would just add or subtract a "1". However to move front to back in the Y axis you would add or subtract "8" for a 8x8x8 cube. To move up or down in the Z axis you select a higher or lower layer.

Just to make sure of the orientation, the X axis is right to left, the Y axis is front to back, and the Z axis is top to bottom. This is illustrated in this next picture.

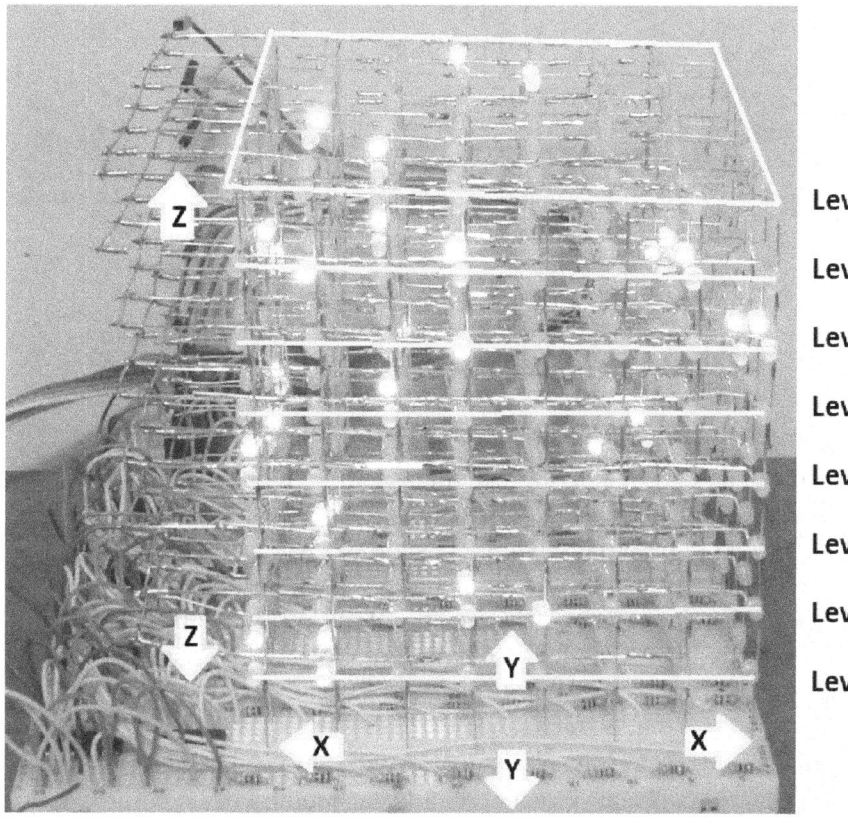

Level 1

Level 2

Level 3

Level 4

Level 5

Level 6

Level 7

Level 8

Before we leave this subject, I want to discus the choice of common anode versus common cathode displays. Common anode displays can not use a ULN2803 driver. There are drivers for common anode displays but usually you will end up using eight driver transistors instead. Since the IC's that are driving the LED's usually have more ability to sink current that to source it, the display on a common anode display would logically be brighter. However I have not seen any difference in the brightness in actual operation.

On a common cathode display a common IC the ULN2803 can be used thus reducing the parts count. But what about the limit on how much current an IC can source? Well if you understand the "guts" of a typical TTL or HCT IC then you know that there is a current limiting resistor built into the output stage of most TTL IC's.

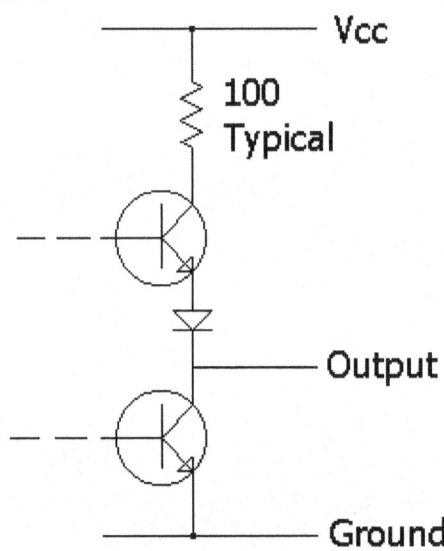

In reality you can actually get away with having no external current limiting resistor! However I would recommend using either 50 or 100 ohms to prevent a meltdown in case there is a short somewhere. Since modern LED's light at around 3.4 volts and a logic high is really only about 4.5 volts then you will have I = E/R of 1/100 = .01 amps or 10 ma. With a 100 ohm resistor. A 50 ohm resistor would then yield a maximum current of .02 amps or 20 ma. That is quite safe!

Another consideration is to use shift registers with built in latches or to use latches. The use of latches may seem like a faster method of updating a display. You can output all eight bits with one output command. However if you have eight latches then you have to turn off the display while the eight latches are being updated. Then once they are updated and the data is loaded you can turn the display back on. That means that as much as half the time the display is not lit up! The final result might be a dimmer display.

If you use shift registers then you have to wait while 64 bits are shifted out to an 8x8 display. However during that time the display will be "on" so that maximum brightness is still achieved. Yes it takes longer to shift out 64 bits to the display. However since the display is working normally during that time, the display "on" time will not be shortened.

So in conclusion shift registers with a latch will yield the brightest display and using a common cathode arrangement will yield the simplest electronic design.

I have included LED cube designs that are both common anode and common cathode. I have also included LED cube designs that are direct drive, shift register drive, and even latch driven (The 8x8x8 cube kit). For cube size I have included 4x4x4 and 8x8x8 arrangements as well as some variations of those. That way most of the possible LED cube configurations are covered in this book.

Chapter 3

4x4x4 Direct Drive Cube

The first project is a 4x4x4 direct drive LED cube. This project does not need any interface IC's but it uses every last pin of the Arduino to directly drive the cube. This leaves zero pins left over for things like a switch or analog input. It also is not recommended for use with the PcDuino as it requires too much current from the I/O pins.

To make the LED cube we must first make four 4x4 LED arrays that are transparent. To do this we will need a wood template with 1/8 inch holes with a one inch by one inch spacing. The 1/8 inch holes are for 3mm LED's. Larger LED's would require larger holes. I used flat top 3mm LED's as I had many of them around that were left over from another project. Diffused 3mm LED's should work just as well if not better.

The one inch spacing requires the use of wires to connect the LED's together. Their leads are not quite long enough to use them for one inch spacing. If the spacing was to be reduced to .8 inches then you could reliably use the leads to connect the LED's together.

To make the transparent LED arrays first take the longer positive lead of the LED's and bend it into a loop about 3/8 inches from the LED. Cut off the excess lead. Then insert the LED's in the holes that you drilled in the wood. Then take a four inch piece of wire and poke it through the loops that you just created. Then solder the wire in place.

One of the tricks to soldering LED's is to never solder next to the LED itself. Always keep the soldering iron at least 1/4 inch away from the LED. You can solder from the farthest point that touches both wires being soldered together. The solder should flow down the wire and make a solid connection.

Coming up next is a picture of some LED's with the longer lead bent into a loop.

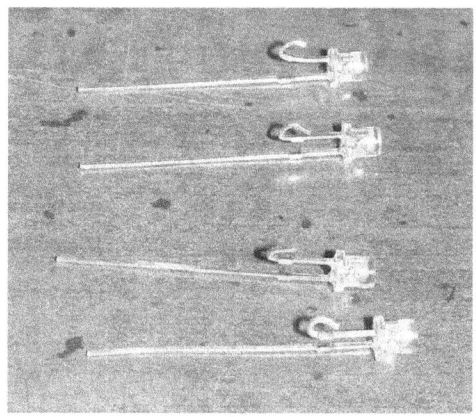

The other lead from the LED is bent flat to the wood board. When all four rows of LED's are in place a second wire coming from the other direction is soldered to the second, negative lead that forms the columns. This is done four times for each of the four columns.

Up next is a picture of the LED's in the wood frame. Note that I marked where to cut the wires by arrows visible on the right, so the wires are all about the same length.

Up next is a close up of the LED matrix frame wiring. The left lead is going to the column wire and the right lead is bent flat and connected to the row wire.

When you have soldered up all 16 LED's you can use a 9 volt battery and a 470 ohm resistor to test out the LED's. Then make any repairs that might be necessary. Once verified to work the LED array can then be removed and then another array can be made. You will need to make four of the 4x4 LED arrays.

The arrays should plug into two breadboards on the one inch centers. You might want to install the 100 ohm resistors first. I cut the leads shorter on the resistors to mount them flat to the breadboard.

Once the arrays are in place you will need to make the cross connectors. I used the same board and inserted round IC socket type connectors in the holes that the LED's were in. These connectors were made by cutting a socket apart into individual pins. Then you solder a wire across each of the four socket pins to make the cross connectors.

On the next page there is a picture of the cross connectors being made. Also one of the IC sockets is visible at the left that has not yet been cut into individual pieces.

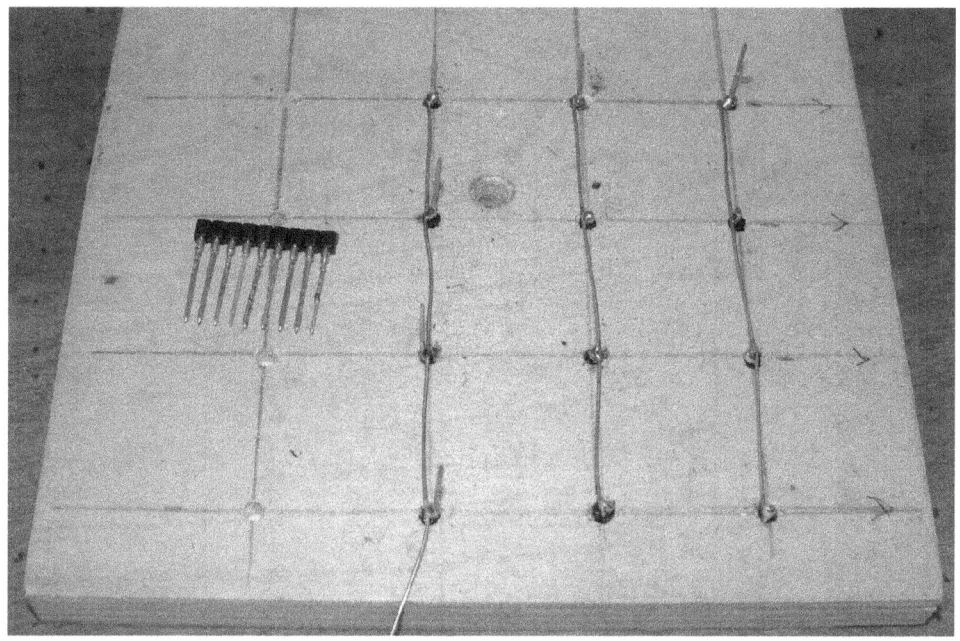

If it is wired up correctly the back row of LED's will correspond to the first line of 1's and 0's in the code. If you do it backwards it works almost as well. Once it is all wired up you can then run a LED test program. With this design if something does not work it is not hard to take it apart and fix the problem

On the next page there is the schematic diagram. It is turned sideways to better fit on the page.

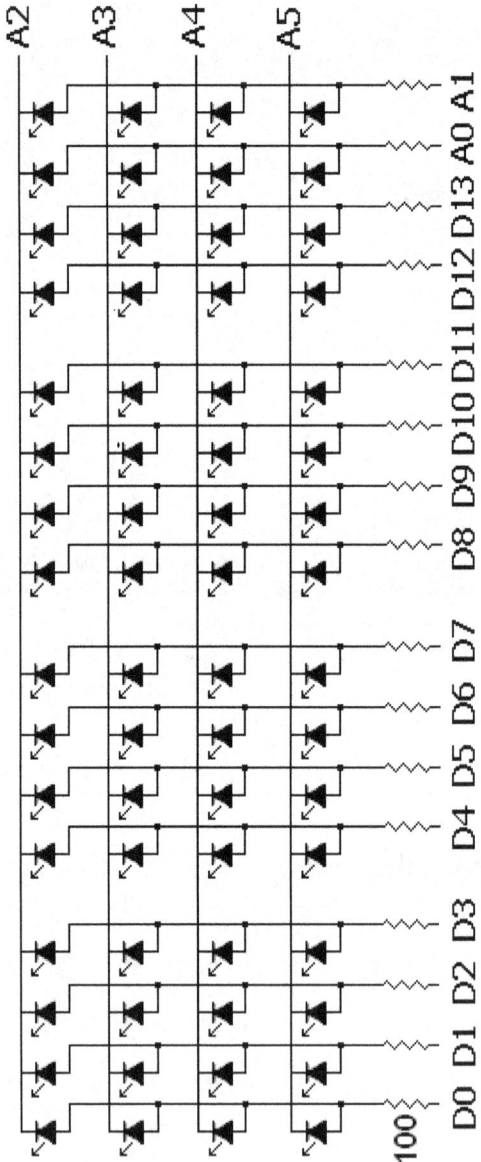

The next picture is a front view of the LED cube with the Arduino Uno that is running it visible on the right side.

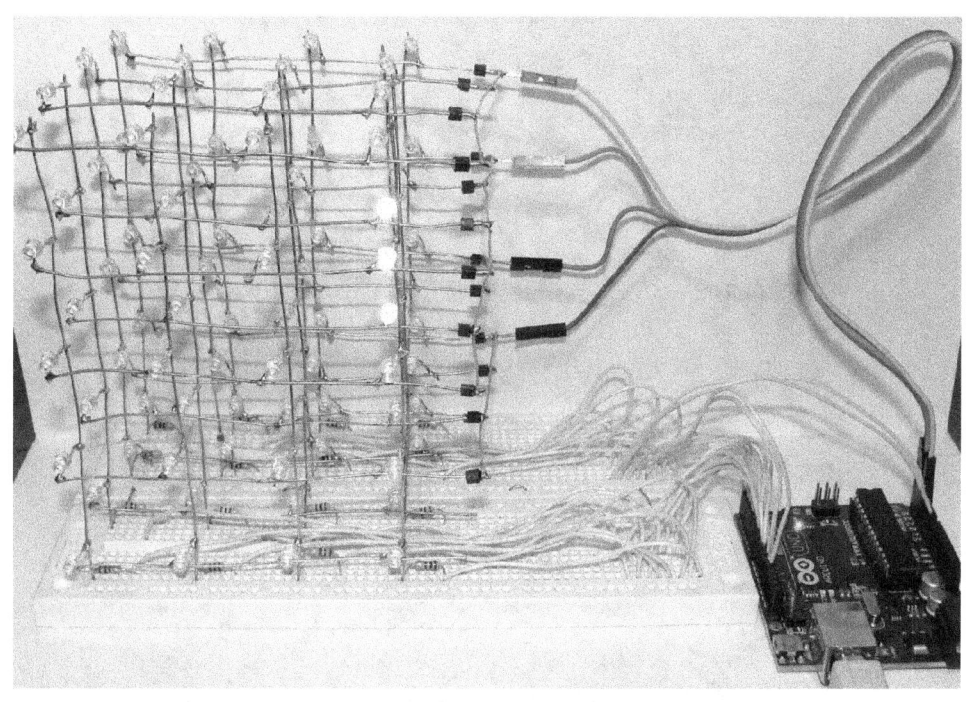

This is a top view of the directly driven LED cube.

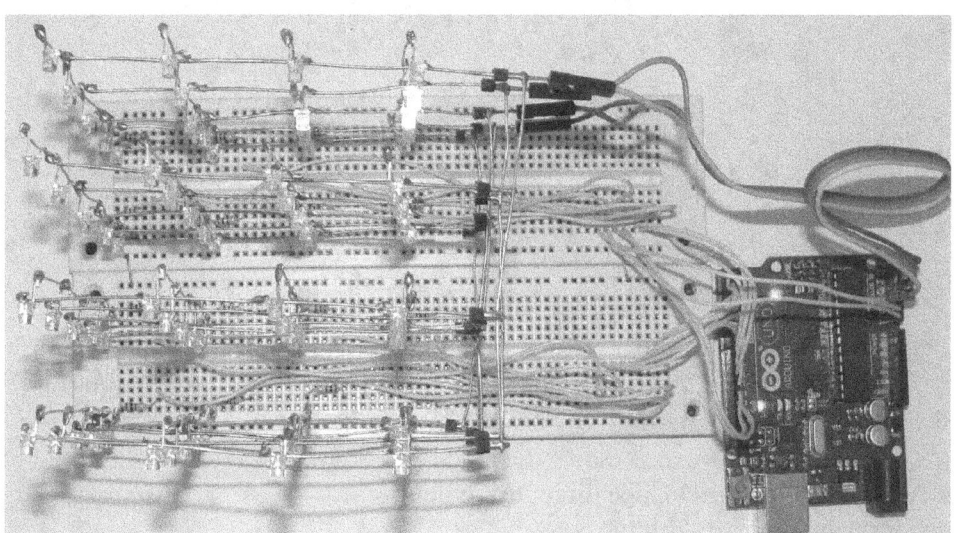

Up next is a side view of the 4x4x4 cube. From this view you can see the 100 ohm resistors and how the arrays can easily be disassembled for repairs or storage.

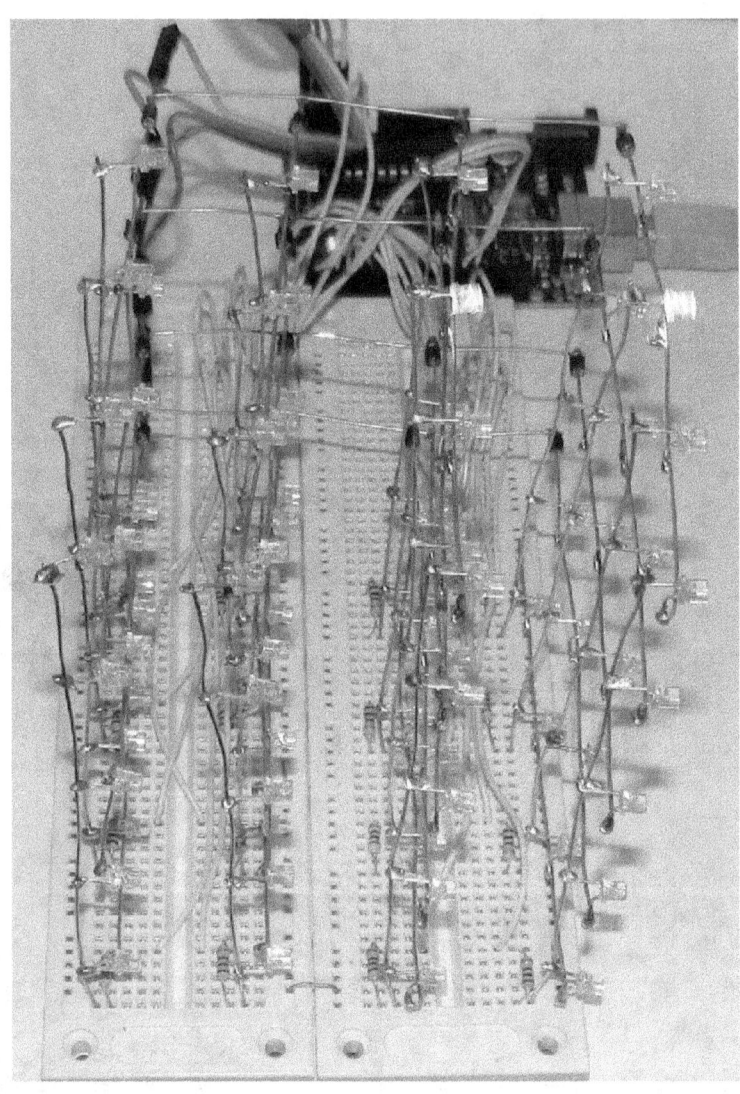

This is a demo program that creates a diagonal plane that rotates up through the cube. The concepts to animate almost anything are demonstrated in this sketch. Feel free to change the 1's and 0's to make your own animations.

```
// Arduino 4x4x4 LED Cube direct drive demo
// store data onboard Arduino
#include <avr/pgmspace.h>
// Define pins connected to cube
int LEDPin[16] = {0, 1, 2, 3, 4, 5, 6, 7, 8, 9, 10, 11, 12, 13, A0, A1};
int LevelPin[4] = {A2, A3, A4, A5};
```

```
// Data arrays
byte Line[][16]= {
 {0,0,0,0,1,0,0,0,0,1,0,0,0,0,1,0},
 {1,0,0,0,0,1,0,0,0,0,1,0,0,0,0,1},
 {0,1,0,0,0,0,1,0,0,0,0,1,0,0,0,0},
 {0,0,1,0,0,0,0,1,0,0,0,0,1,0,0,0},
 {0,0,0,1,0,0,0,0,1,0,0,0,0,1,0,0},
 {0,0,0,0,1,0,0,0,0,1,0,0,0,0,1,0},
 {1,0,0,0,0,1,0,0,0,0,1,0,0,0,0,1},
 {0,1,0,0,0,0,1,0,0,0,0,1,0,0,0,0},
 {0,0,1,0,0,0,0,1,0,0,0,0,1,0,0,0},
 };

// Setup Arduino pins
void setup(){
  // Set 16 LED pins as outputs HIGH=lit
  for (int pin=0; pin < 16; pin++) {
   pinMode( LEDPin[pin], OUTPUT );
  }
  // Set Level pins as outputs LOW=lit
  for (int pin=0; pin < 4; pin++) {
   pinMode( LevelPin[pin], OUTPUT );
  }
}
int shift = 0;
int cycle = 0;
// Start the program
void loop(){
  for (int led=0; led < 16; led++){
   if (Line[0+shift][led] == 1){
    digitalWrite(LEDPin[led],HIGH);
   } else {
    digitalWrite(LEDPin[led],LOW);
  } }
  digitalWrite(A2,LOW);
  delay(5);
  digitalWrite(A2,HIGH);
  for (int led=0; led < 16; led++){
   if (Line[1+shift][led] == 1){
    digitalWrite(LEDPin[led],HIGH);
   } else {
```

```
      digitalWrite(LEDPin[led],LOW);
  } }
digitalWrite(A3,LOW);
delay(5);
digitalWrite(A3,HIGH);
for (int led=0; led < 16; led++){
  if (Line[2+shift][led] == 1){
    digitalWrite(LEDPin[led],HIGH);
  } else {
    digitalWrite(LEDPin[led],LOW);
  } }
digitalWrite(A4,LOW);
delay(5);
digitalWrite(A4,HIGH);
for (int led=0; led < 16; led++){
  if (Line[3+shift][led] == 1){
    digitalWrite(LEDPin[led],HIGH);
  } else {
    digitalWrite(LEDPin[led],LOW);
  } }
digitalWrite(A5,LOW);
delay(5);
digitalWrite(A5,HIGH);
cycle = cycle+1;
if (cycle == 25) {
  shift = shift+1;
  cycle = 0;
 }
if (shift > 4) shift=0;
}
```

Chapter 4

4x4x5 LED Cube with 74138

Each project in this book gets a little bit more complex. To make this cube we must first make four 4x5 LED arrays that are transparent. Use the same template as was used for the 4x4x4 cube but add another row one inch below the other rows. You can make the 4x5 arrays from scratch or you can add a fifth row onto the 4x4 LED arrays that were used in the previous project.

After you have added the fifth row and remounted the LED arrays, add the 74138 IC and connect it up. The 74138 is a three line of eight line decoder. This chip can take the binary values of 0 to 7 (111) and from that value select one of eight outputs (0 to 7). This improved design leaves a spare analog input pin. It also supports having up to eight LED layers!

Once it is all wired up run a LED test program. Hopefully you will only damage one or two LED's and they will need to be replaced if they do not work.

This setup is less than perfect because the routine to select the level takes too long to complete. Since the display is on as each of the three bits are sent to the 74138 to select the level it wit displays a "preheat" in nearby levels. That is to say that some LED's will glow dimly before they are lit up completely.

Another problem is that the 74LS138 (or 74HCT138) cannot deliver quite enough power. I actually "piggy backed" a second 74LS138 on top of the first 74LS138 to get a little bit more brightness.

Up next is the schematic diagram of how to add the 74138. I actually used a 74HCT138.

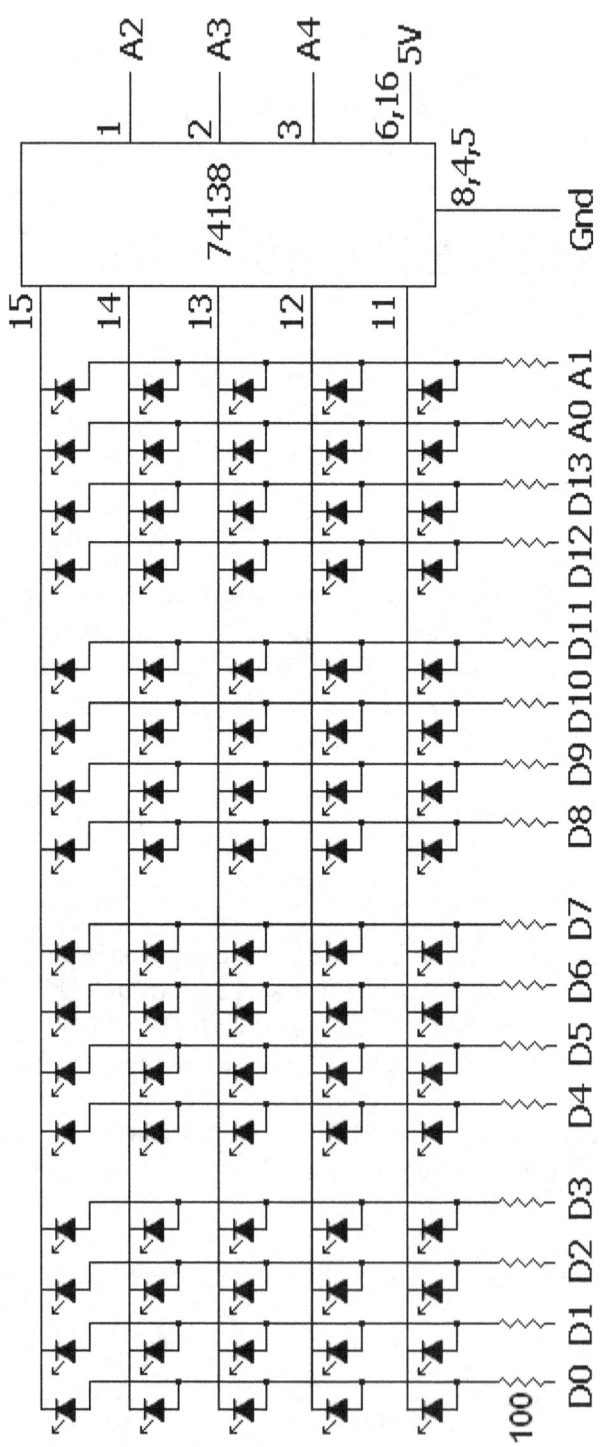

Here is a picture of the LED cube with 74138 that was taken from above.

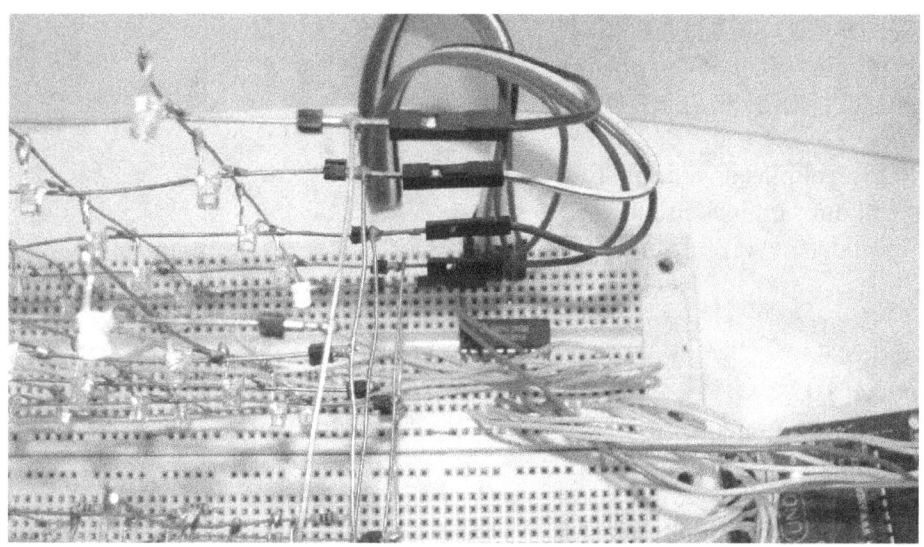

Here is the same demo program but this time it is set up for the 74138. It has also been optimized so that each level reuses the same loop thus making it much shorter.

```
// Arduino 4x4x4 LED Cube with 74138 demo
// store data onboard Arduino
#include <avr/pgmspace.h>
// Defining pins connected to cube
int LEDPin[16] = {0, 1, 2, 3, 4, 5, 6, 7, 8, 9, 10, 11, 12, 13, A0, A1};
int LevelPin[4] = {A2, A3, A4};

// Data arrays
byte Line[][16]= {
{0,0,0,0,1,0,0,0,0,1,0,0,0,0,1,0},
{1,0,0,0,0,1,0,0,0,0,1,0,0,0,0,1},
{0,1,0,0,0,0,1,0,0,0,0,1,0,0,0,0},
{0,0,1,0,0,0,0,1,0,0,0,0,1,0,0,0},
{0,0,0,1,0,0,0,0,1,0,0,0,0,1,0,0},
{0,0,0,0,1,0,0,0,0,1,0,0,0,0,1,0},
{1,0,0,0,0,1,0,0,0,0,1,0,0,0,0,1},
{0,1,0,0,0,0,1,0,0,0,0,1,0,0,0,0},
{0,0,1,0,0,0,0,1,0,0,0,0,1,0,0,0},
};
```

```
// Setup Arduino pins
void setup(){
  // Set 16 LED pins as outputs HIGH=lit
  for (int pin=0; pin < 16; pin++) {
    pinMode( LEDPin[pin], OUTPUT );
  }
  // Set Level pins as outputs LOW=lit
  for (int pin=0; pin < 3; pin++) {
    pinMode( LevelPin[pin], OUTPUT );
  }
}
int shift = 0;
int cycle = 0;
// Start the program
void loop(){
  // Turn off display
  digitalWrite(A2,HIGH);
  digitalWrite(A3,HIGH);
  digitalWrite(A4,HIGH);
  for (int level=0; level < 5; level++){
    for (int led=0; led < 16; led++){
      if (Line[level+shift][led] == 1){
        digitalWrite(LEDPin[led],HIGH);
      } else {
        digitalWrite(LEDPin[led],LOW);
    } }
    if bitRead(level, 0){
      digitalWrite(A2,HIGH);
    } else {
      digitalWrite(A2,LOW);
    }
    if bitRead(level, 1){
      digitalWrite(A3,HIGH);
    } else {
      digitalWrite(A3,LOW);
    }
    if bitRead(level, 2){
      digitalWrite(A4,HIGH);
    } else {
      digitalWrite(A4,LOW);
    }
```

```
    delay(3);
  }
  cycle = cycle+1;
  // Rate of movement - one per 25 cycles
  if (cycle == 25) {
    shift = shift+1;
    cycle = 0;
  }
  if (shift > 4) shift=0;
}
```

Chapter 5

4x4x5 LED Cube with 74595's

This project also uses four of the 4x5 LED arrays. They follow the same design as the 4x4x4 cube LED arrays. Just add another row following the same pattern.

The added 74595 IC is a shift register with a latch on the outputs. That latch allows you to shift data in while it is displaying the old data. Then with the register clock command the latch is updated showing the new data. This prevents any "ghosts" or "preheat" as I called them in the last project.

I connected pin 7 of the back 74HC595 to the back left corner of the LED arrays. Then connect the other pins following in that order. Pin 15 of the front 74595 should end up going to the right front corner of the array. If it is done right the back row of LED's will correspond to the first line of 1's and 0's in the code.

There is also a ULN2003 driver IC added to help select the levels. The ULN2003 can sink 1/2 an amp. The Arduino is rated for 40 ma and when 16 LED's are on, at 10 ma each, that is 160 ma. That is four times the rated power of the Arduino Uno.

Coming up next is a picture of the LED cube with the 74595's and ULN2003 IC's added. The two shift registers are at the right end of the breadboard. The ULN2003 is located under the cathode or negative connection pins.

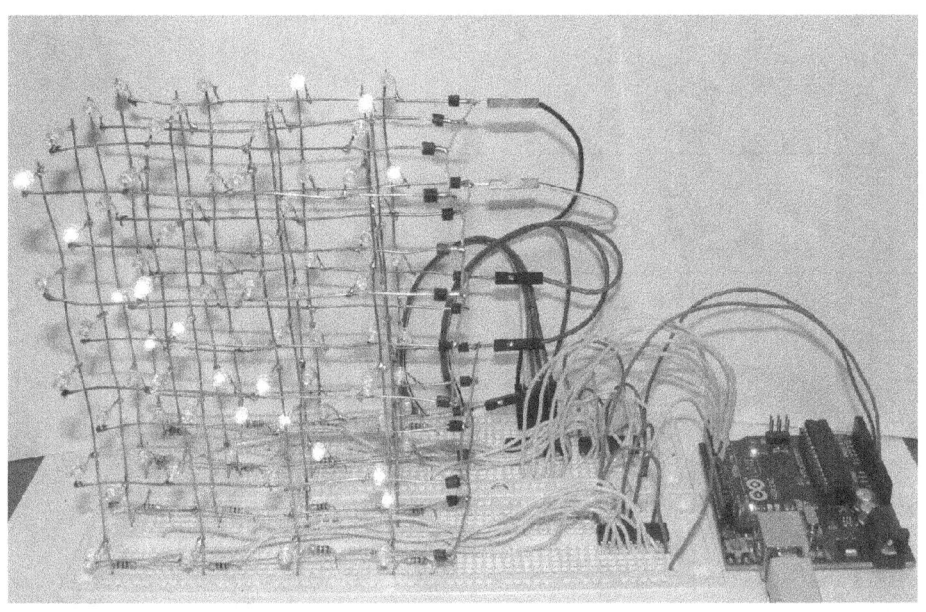

Here is a picture from above showing the three IC's on the right.

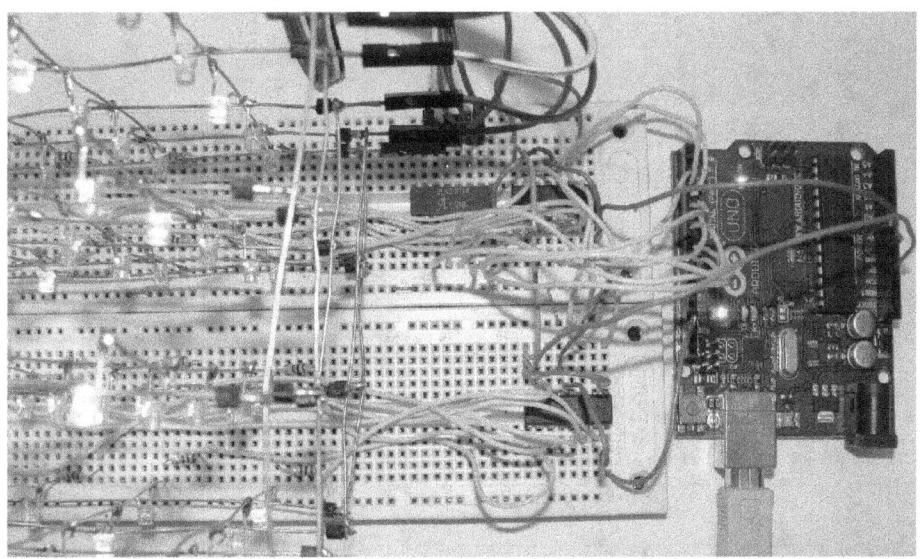

Coming up next is the schematic diagram of the 4x4x5 LED cube with 74595's. It is once again turned sideways to fit it in this book. The schematic makes it look simpler that it really is to make.

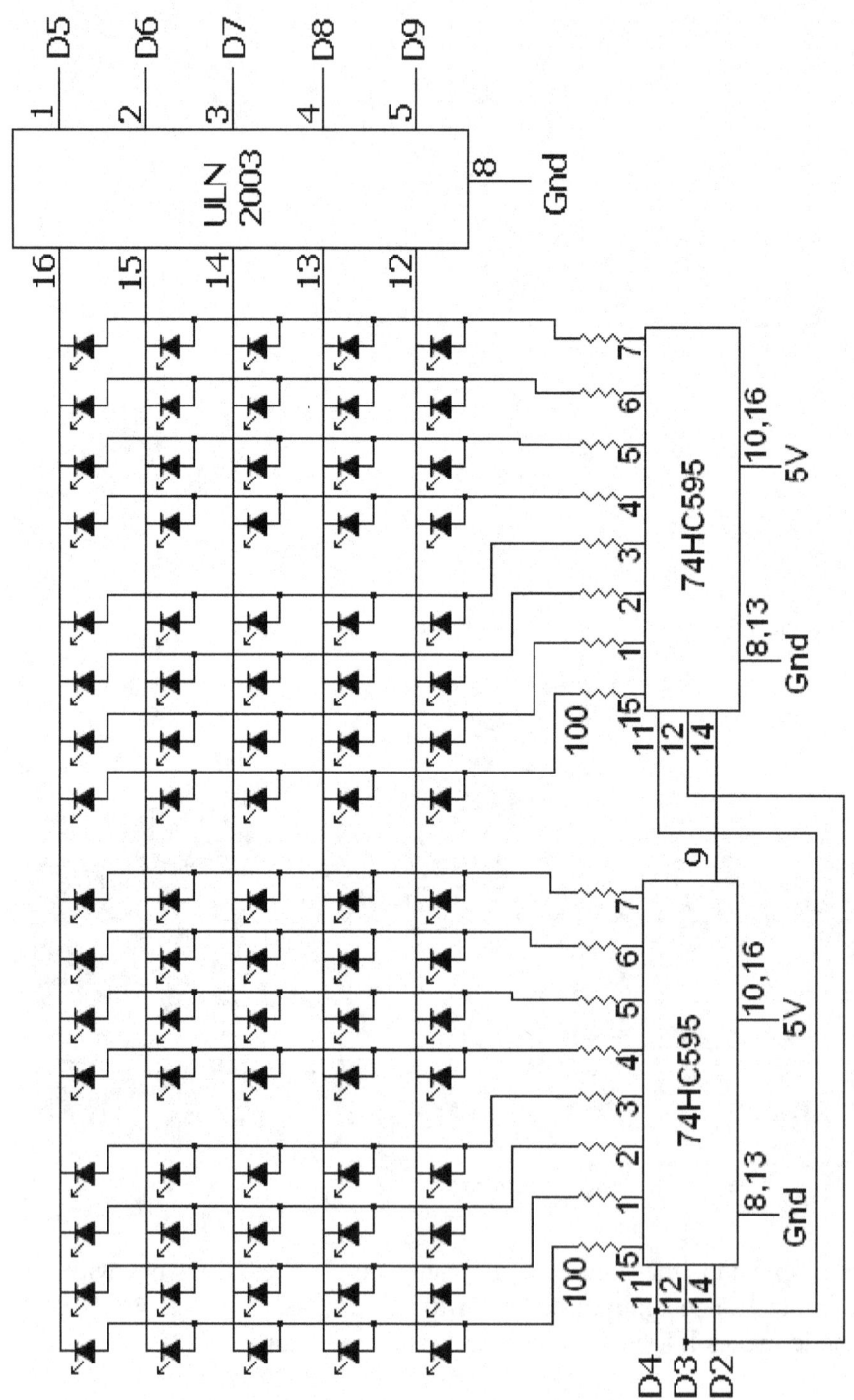

28

Here is the same demo program as before but this time it was re-written for the 74595 shift registers.

```
// LED Cube with 74595's and ULN2003 demo
// These Pins Connect to 74595's
int data = 2;
int clock = 3;
int latch = 4;
// These Pins Connect to ULN2003
int Level1 = 5;
int Level2 = 6;
int Level3 = 7;
int Level4 = 8;
int Level5 = 9;

int rotate = 0;
int cycle = 0;
byte Line[][16]= {
// {1,1,1,1,1,1,1,1,1,1,1,1,1,1,1,1},
  {1,0,0,0,0,1,0,0,0,0,1,0,0,0,0,1},
  {0,1,0,0,0,0,1,0,0,0,0,1,0,0,0,0},
  {0,0,1,0,0,0,0,1,0,0,0,0,1,0,0,0},
  {0,0,0,1,0,0,0,0,1,0,0,0,0,1,0,0},
  {0,0,0,0,1,0,0,0,0,1,0,0,0,0,1,0},
  {1,0,0,0,0,1,0,0,0,0,1,0,0,0,0,1},
  {0,1,0,0,0,0,1,0,0,0,0,1,0,0,0,0},
  {0,0,1,0,0,0,0,1,0,0,0,0,1,0,0,0},
};

// set up output pins
void setup() {
  pinMode(data, OUTPUT);
  pinMode(clock, OUTPUT);
  pinMode(latch, OUTPUT);

  pinMode(Level1, OUTPUT);
  pinMode(Level2, OUTPUT);
  pinMode(Level3, OUTPUT);
  pinMode(Level4, OUTPUT);
  pinMode(Level5, OUTPUT);
}
```

```
void loop() {
  for (int level=0; level <5; level++){
    for (int shift=0; shift <16; shift++){
      if (Line[level+rotate][shift] == 1){
        digitalWrite(data, HIGH);
      } else {
        digitalWrite(data, LOW);
      }
      // Clocks in the new data
      digitalWrite(clock, LOW);
      digitalWrite(clock, HIGH);
    }
    // Turn the levels off
    digitalWrite(Level1, LOW);
    digitalWrite(Level2, LOW);
    digitalWrite(Level3, LOW);
    digitalWrite(Level4, LOW);
    digitalWrite(Level5, LOW);
    //Latches in the new data
    digitalWrite(latch, LOW);
    digitalWrite(latch, HIGH);
    // Select the new level to turn on
    if (level==0)digitalWrite(Level1, HIGH);
    if (level==1)digitalWrite(Level2, HIGH);
    if (level==2)digitalWrite(Level3, HIGH);
    if (level==3)digitalWrite(Level4, HIGH);
    if (level==4)digitalWrite(Level5, HIGH);
    delay(3);
  }
  // Rate of movement - one per 25 cycles
  cycle=cycle+1;
  if (cycle == 25) {
    rotate = rotate+1;
    cycle = 0;
  }
  if (rotate > 4) rotate=0;
}
```

You can now use the analog inputs to animate the cube. With just a few changes to the loop this program will make each layer take a sample from analog input 0 and then display the results. For testing take the sound from something like a laptop or mp3 player and connect it to ground and analog input 0. Make sure that it does not exceed 5 volts!

```
// LED Cube with 74595's and ULN2003 sound demo

// These Pins Connect to 74595's
int data = 2;
int clock = 3;
int latch = 4;
// These Pins Connect to ULN2003
int Level1 = 5;
int Level2 = 6;
int Level3 = 7;
int Level4 = 8;
int Level5 = 9;

int rotate = 0;
int cycle = 0;

// set up output pins
void setup() {
  pinMode(data, OUTPUT);
  pinMode(clock, OUTPUT);
  pinMode(latch, OUTPUT);
  pinMode(Level1, OUTPUT);
  pinMode(Level2, OUTPUT);
  pinMode(Level3, OUTPUT);
  pinMode(Level4, OUTPUT);
  pinMode(Level5, OUTPUT);
}

void loop() {
  for (int level=0; level <5; level++){
    // collect audio samples
    int input1=analogRead(0)/16;
    for (int shift=0; shift <16; shift++){
      if (shift < input1){
        digitalWrite(data, HIGH);
      } else {
        digitalWrite(data, LOW);
```

```
  }
  // Clocks in the new data
  digitalWrite(clock, LOW);
  digitalWrite(clock, HIGH);
}
// Turn the levels off
digitalWrite(Level1, LOW);
digitalWrite(Level2, LOW);
digitalWrite(Level3, LOW);
digitalWrite(Level4, LOW);
digitalWrite(Level5, LOW);
//Latches in the new data
digitalWrite(latch, LOW);
digitalWrite(latch, HIGH);
// Select the new level to turn on
if (level==0)digitalWrite(Level1, HIGH);
if (level==1)digitalWrite(Level2, HIGH);
if (level==2)digitalWrite(Level3, HIGH);
if (level==3)digitalWrite(Level4, HIGH);
if (level==4)digitalWrite(Level5, HIGH);
delay(3);
}
}
```

Chapter 6

8x8x8 LED Cube

For this next project we are going to make a cube that is four times as complex. First the spacing between the LED's was reduced from one inch to .6 inches. That way the LED leads can be soldered between the LED's without adding any extra wires. In fact I cut about ¼ inch off the longer lead.

First you will need a piece of wood at least five inches by five inches in size. Lay out an 8x8 cross grid that is at 6/10 of an inch intervals. If you go any bigger then you will need to mount the shift registers on a separate breadboard. This spacing is also compatible with a common 8x8x8 LED kit as seen on Instructables. The kit says that the LED spacing is 1.5 cm. Drill 1/8 inch diameter holes for 3mm LED's.

On each LED bend the shorter or negative lead at about 1/4 inch or just above the funny spot in the lead. Put 8 of them into the template. Bend the long leads down in the other direction, and solder them together. Then do the same for the next 8 LED's but this time also solder the shorter leads together.

On the next page there is a picture of the board with the first row of LED's in place.

The next picture shows four rows that have been completed. If you install the LED's early you will have to remove them in order to bend their leads.

Here is another close up view from the other direction so you can see how the LED's are connected. The negative, shorter leads are about to be soldered.

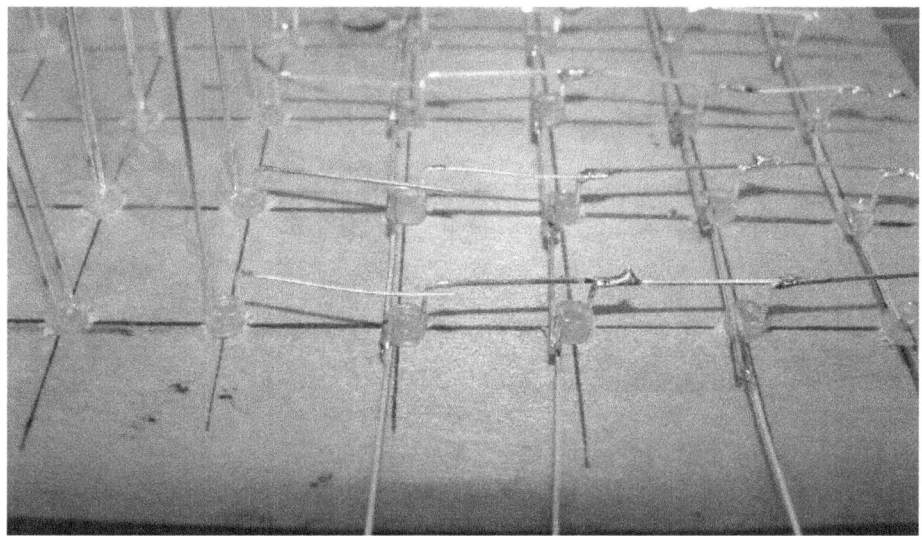

The next picture is of a finished 8x8 LED array. Once all 64 LED's are soldered test them out with a 9 volt battery and a 470 ohm resistor. Also it is a bit tricky to remove the array from the template. You have to gently pull up on each section until it finally lifts out of the wood template.

You will also need to modify the breadboards for this project. The tighter spacing requires that the breadboards be closer together than one inch. To accomplish that we will need to remove the power strips from both sides of the breadboard. These instructions apply to the MB-102 as well as many other breadboards. You will need four of these breadboards to fit the eight 8x8 LED arrays at two per breadboard.

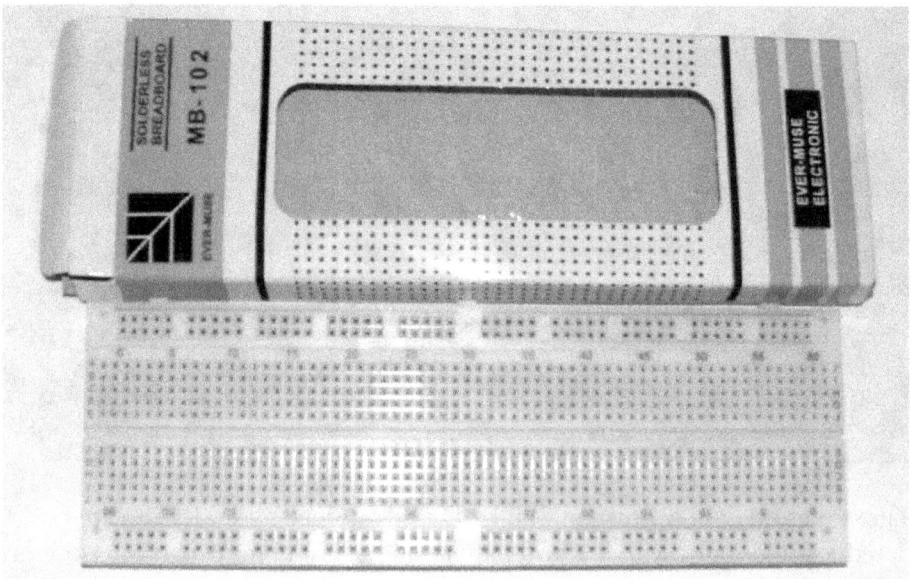

Flip the breadboard over and cut the backing 3/8 of an inch in from each side. The power strips then should either slide up or down to remove them.

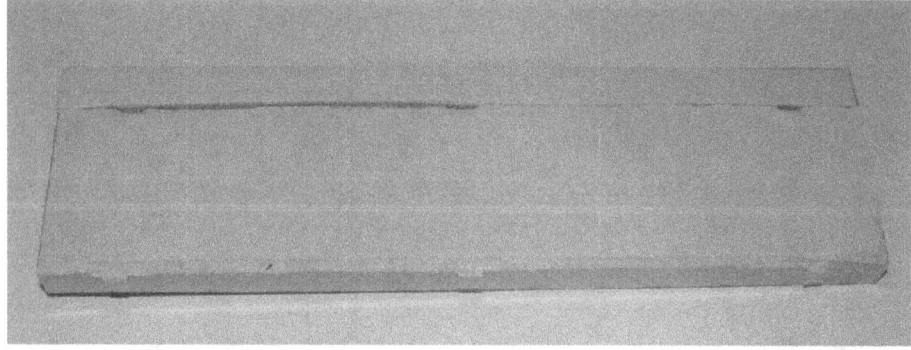

Once the power strips are removed you can usually connect them together. The spacing should now be about .7 inches. That is a little more than the .6 inch spacing between the LED's but it will work. You can now start adding the LED arrays, adding the two 74HC595's, and then the 100 ohm resistors. Up next there is a picture that shows what you should have assembled so far.

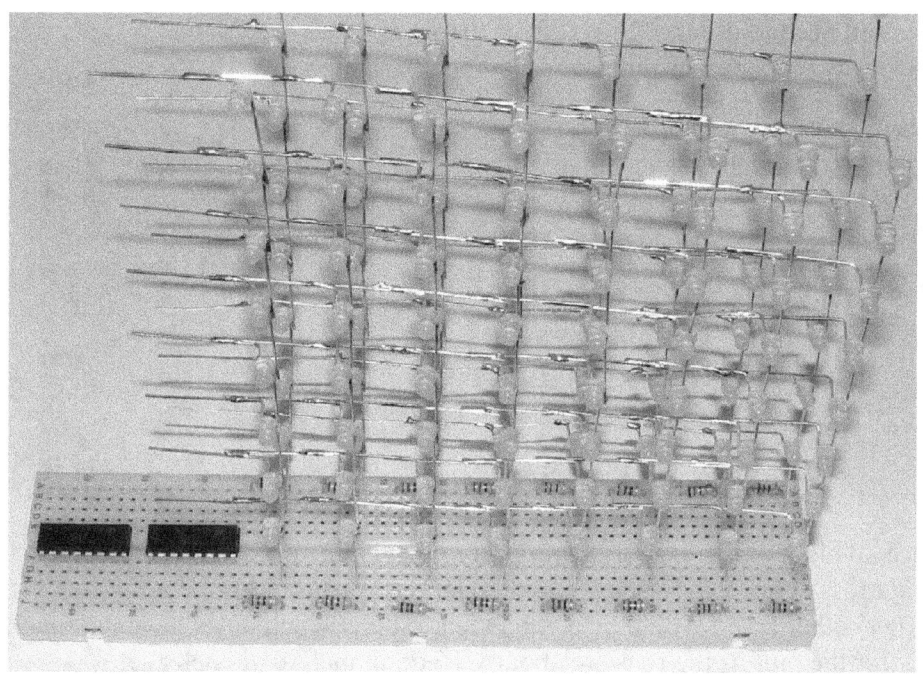

Next you add the 16 jumper wires to the resistors. You will need some two inch, four inch and six inch jumpers. The furthest resistor goes to pin 15, then the next goes to pin one, etc. The right IC goes to the back array and the left IC goes to the front array. Up next is a picture with the 16 jumpers installed.

Note that the centers of the arrays need to be slightly offset on the outside. The breadboards are at .7 inch centers and the arrays are only at .6 inches.

You will need to use row E, I, C, H, C, H, B and F as seen in the next picture. One of my breadboards did not match the other three.

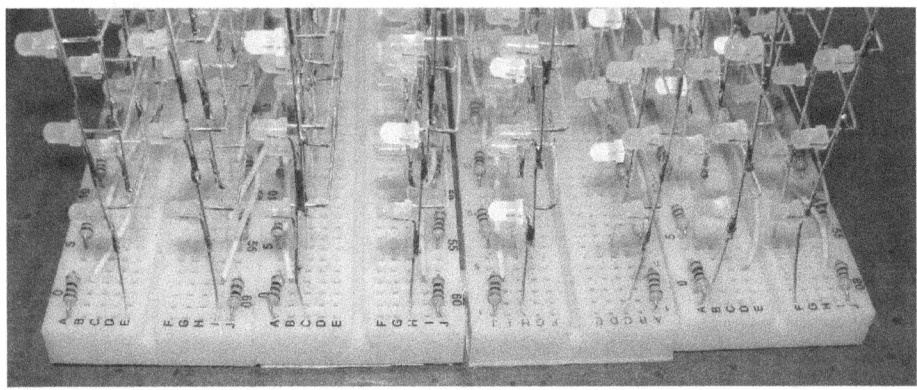

Once all eight of the 8x8 LED arrays are made then you will need to make eight cross connects. I used the same board as was used for the arrays but using another board would be a good idea. The 1/8 inch holes have too much play for the pins as the pins are about 1/16 of an inch in diameter. It was really hard to get the pins straight in the larger holes. I included an IC socket and some salvaged pins on the right side of this picture. To get the pins out of the socket cut the socket close to the pins and then pry the pin out of the socket.

Coming up next is a picture of the completed 8x8x8 LED cube running a LED test. The wiring is fairly difficult for the inner IC's as they are located underneath of the array interconnects. If you connect the wire from pin 14 of the first 74HCT595 that normally goes to the Arduino D2 to 5 volts then all of the LED's should come on to test them to see if they all work.

The next picture is a top view of the completed cube. In this picture you can see the Arduino Uno and ULN2003 off to the left side. Also the shift registers are more visible in this picture. You can also see that the LED arrays are only fastened in place on the left side and bottom.

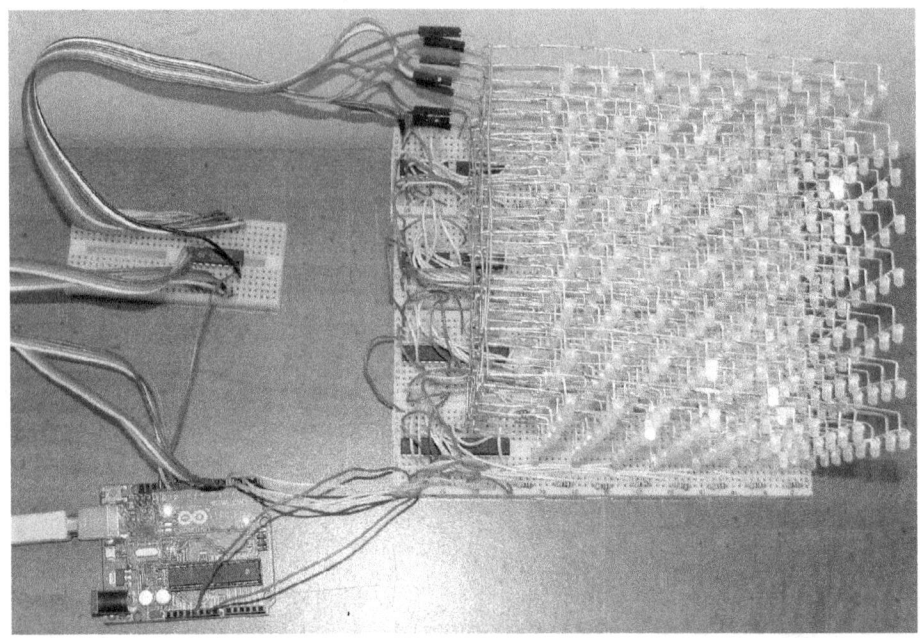

On the next page there is the schematic diagram for two of the eight cube layers. The ULN2803 IC is shared among all eight of the cube's layers. Each 8x8 LED array has its own 74595 shift register. So on each breadboard there should be two arrays on the right and two shift registers on the left. Four modules like what is shown in the schematic diagram are made, then the four modules are connected together to make the 8x8x8 LED cube.

Note that if all 64 of the LED's are lit at the same time at 10 ma each you would have 640 ma and thus exceed the ULN2803 power rating. This did not cause any problems in any of my tests, but it is very rare to have all of the LED's on at the same time.

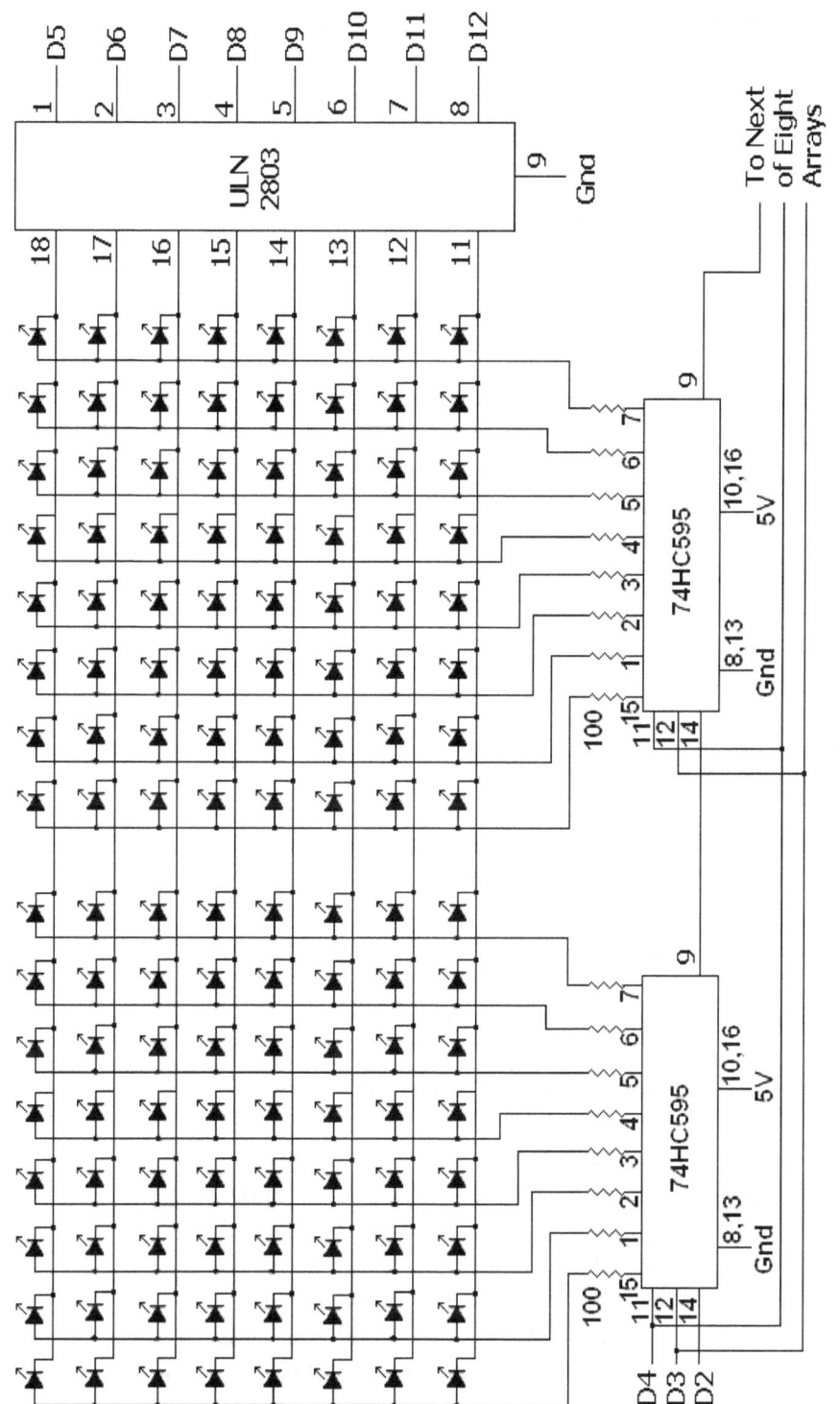

Here is the code for the diagonal slice program that works just like the program that was used for the 4x4x5 LED cube. This time we will need arrays of 64 bits per level instead of 16.

```
// 8x8x8 LED Cube diagonal slice demo

// These Pins Connect to 74595's
int data = 2;
int clock = 4;
int latch = 3;
// These Pins Connect to ULN2003
int Level1 = 5;
int Level2 = 6;
int Level3 = 7;
int Level4 = 8;
int Level5 = 9;
int Level6 = 10;
int Level7 = 11;
int Level8 = 12;

int rotate = 0;
int cycle = 0;
byte Line[][64]= {
// there are eight groups of eight LED's per layer.
{1,0,0,0,0,0,0,0, 0,1,0,0,0,0,0,0, 0,0,1,0,0,0,0,0, 0,0,0,1,0,0,0,0,
0,0,0,0,1,0,0,0, 0,0,0,0,0,1,0,0, 0,0,0,0,0,0,1,0, 0,0,0,0,0,0,0,1},
 {0,1,0,0,0,0,0,0, 0,0,1,0,0,0,0,0, 0,0,0,1,0,0,0,0, 0,0,0,0,1,0,0,0,
0,0,0,0,0,1,0,0, 0,0,0,0,0,0,1,0, 0,0,0,0,0,0,0,1, 1,0,0,0,0,0,0,0},
 {0,0,1,0,0,0,0,0, 0,0,0,1,0,0,0,0, 0,0,0,0,1,0,0,0, 0,0,0,0,0,1,0,0,
0,0,0,0,0,0,1,0, 0,0,0,0,0,0,0,1, 1,0,0,0,0,0,0,0, 0,1,0,0,0,0,0,0},
 {0,0,0,1,0,0,0,0, 0,0,0,0,1,0,0,0, 0,0,0,0,0,1,0,0, 0,0,0,0,0,0,1,0,
0,0,0,0,0,0,0,1, 1,0,0,0,0,0,0,0, 0,1,0,0,0,0,0,0, 0,0,1,0,0,0,0,0},
 {0,0,0,0,1,0,0,0, 0,0,0,0,0,1,0,0, 0,0,0,0,0,0,1,0, 0,0,0,0,0,0,0,1,
1,0,0,0,0,0,0,0, 0,1,0,0,0,0,0,0, 0,0,1,0,0,0,0,0, 0,0,0,1,0,0,0,0},
 {0,0,0,0,0,1,0,0, 0,0,0,0,0,0,1,0, 0,0,0,0,0,0,0,1, 1,0,0,0,0,0,0,0,
0,1,0,0,0,0,0,0, 0,0,1,0,0,0,0,0, 0,0,0,1,0,0,0,0, 0,0,0,0,1,0,0,0},
 {0,0,0,0,0,0,1,0, 0,0,0,0,0,0,0,1, 1,0,0,0,0,0,0,0, 0,1,0,0,0,0,0,0,
0,0,1,0,0,0,0,0, 0,0,0,1,0,0,0,0, 0,0,0,0,1,0,0,0, 0,0,0,0,0,1,0,0},
 {0,0,0,0,0,0,0,1, 1,0,0,0,0,0,0,0, 0,1,0,0,0,0,0,0, 0,0,1,0,0,0,0,0,
0,0,0,1,0,0,0,0, 0,0,0,0,1,0,0,0, 0,0,0,0,0,1,0,0, 0,0,0,0,0,0,1,0},
```

```
{1,0,0,0,0,0,0,0, 0,1,0,0,0,0,0,0, 0,0,1,0,0,0,0,0, 0,0,0,1,0,0,0,0,
0,0,0,0,1,0,0,0, 0,0,0,0,0,1,0,0, 0,0,0,0,0,0,1,0, 0,0,0,0,0,0,0,1},
{0,1,0,0,0,0,0,0, 0,0,1,0,0,0,0,0, 0,0,0,1,0,0,0,0, 0,0,0,0,1,0,0,0,
0,0,0,0,0,1,0,0, 0,0,0,0,0,0,1,0, 0,0,0,0,0,0,0,1, 1,0,0,0,0,0,0,0},
{0,0,1,0,0,0,0,0, 0,0,0,1,0,0,0,0, 0,0,0,0,1,0,0,0, 0,0,0,0,0,1,0,0,
0,0,0,0,0,0,1,0, 0,0,0,0,0,0,0,1, 1,0,0,0,0,0,0,0, 0,1,0,0,0,0,0,0},
{0,0,0,1,0,0,0,0, 0,0,0,0,1,0,0,0, 0,0,0,0,0,1,0,0, 0,0,0,0,0,0,1,0,
0,0,0,0,0,0,0,1, 1,0,0,0,0,0,0,0, 0,1,0,0,0,0,0,0, 0,0,1,0,0,0,0,0},
{0,0,0,0,1,0,0,0, 0,0,0,0,0,1,0,0, 0,0,0,0,0,0,1,0, 0,0,0,0,0,0,0,1,
1,0,0,0,0,0,0,0, 0,1,0,0,0,0,0,0, 0,0,1,0,0,0,0,0, 0,0,0,1,0,0,0,0},
{0,0,0,0,0,1,0,0, 0,0,0,0,0,0,1,0, 0,0,0,0,0,0,0,1, 1,0,0,0,0,0,0,0,
0,1,0,0,0,0,0,0, 0,0,1,0,0,0,0,0, 0,0,0,1,0,0,0,0, 0,0,0,0,1,0,0,0},
{0,0,0,0,0,0,1,0, 0,0,0,0,0,0,0,1, 1,0,0,0,0,0,0,0, 0,1,0,0,0,0,0,0,
0,0,1,0,0,0,0,0, 0,0,0,1,0,0,0,0, 0,0,0,0,1,0,0,0, 0,0,0,0,0,1,0,0},
{0,0,0,0,0,0,0,1, 1,0,0,0,0,0,0,0, 0,1,0,0,0,0,0,0, 0,0,1,0,0,0,0,0,
0,0,0,1,0,0,0,0, 0,0,0,0,1,0,0,0, 0,0,0,0,0,1,0,0, 0,0,0,0,0,0,1,0},
};

// set up output pins
void setup() {
  pinMode(data, OUTPUT);
  pinMode(clock, OUTPUT);
  pinMode(latch, OUTPUT);
  pinMode(Level1, OUTPUT);
  pinMode(Level2, OUTPUT);
  pinMode(Level3, OUTPUT);
  pinMode(Level4, OUTPUT);
  pinMode(Level5, OUTPUT);
  pinMode(Level6, OUTPUT);
  pinMode(Level7, OUTPUT);
  pinMode(Level8, OUTPUT);
}

void loop() {
  for (int level=0; level <8; level++){
    for (int shift=0; shift <64; shift++){
     if (Line[level+rotate][shift] == 1){
      digitalWrite(data, HIGH);
     } else {
      digitalWrite(data, LOW);
     }
```

```
  // Clocks in the new data
  digitalWrite(clock, LOW);
  digitalWrite(clock, HIGH);
  }
// Turn the levels off
digitalWrite(Level1, LOW);
digitalWrite(Level2, LOW);
digitalWrite(Level3, LOW);
digitalWrite(Level4, LOW);
digitalWrite(Level5, LOW);
digitalWrite(Level6, LOW);
digitalWrite(Level7, LOW);
digitalWrite(Level8, LOW);
 //Latches in the new data
digitalWrite(latch, LOW);
digitalWrite(latch, HIGH);
// Select the new level to turn on
if (level==0)digitalWrite(Level1, HIGH);
if (level==1)digitalWrite(Level2, HIGH);
if (level==2)digitalWrite(Level3, HIGH);
if (level==3)digitalWrite(Level4, HIGH);
if (level==4)digitalWrite(Level5, HIGH);
if (level==5)digitalWrite(Level6, HIGH);
if (level==6)digitalWrite(Level7, HIGH);
if (level==7)digitalWrite(Level8, HIGH);
  delay(1);
}
// Rate of movement - one per 25 cycles
cycle=cycle+1;
if (cycle == 25) {
  rotate = rotate+1;
 cycle = 0;
 }
 if (rotate > 7) rotate=0;
}
```

With just two changes, this program will create a random, "fireworks" like display. Basically each layer has a random LED that is turned on. You do not need all of the 64 bit data strings that are in the Line array for this program.

```
// 8x8x8 LED Cube random blink demo

// These Pins Connect to 74595's
int data = 2;
int clock = 4;
int latch = 3;
// These Pins Connect to ULN2003
int Level1 = 5;
int Level2 = 6;
int Level3 = 7;
int Level4 = 8;
int Level5 = 9;
int Level6 = 10;
int Level7 = 11;
int Level8 = 12;

int rotate = 0;
int cycle = 0;

// set up output pins
void setup() {
  pinMode(data, OUTPUT);
  pinMode(clock, OUTPUT);
  pinMode(latch, OUTPUT);
  pinMode(Level1, OUTPUT);
  pinMode(Level2, OUTPUT);
  pinMode(Level3, OUTPUT);
  pinMode(Level4, OUTPUT);
  pinMode(Level5, OUTPUT);
  pinMode(Level6, OUTPUT);
  pinMode(Level7, OUTPUT);
  pinMode(Level8, OUTPUT);
}

void loop() {
  for (int level=0; level <8; level++){
    int rdata = random(64);
    for (int shift=0; shift <64; shift++){
      if (rdata == shift){
        digitalWrite(data, HIGH);
      } else {
```

```
      digitalWrite(data, LOW);
    }
    // Clocks in the new data
    digitalWrite(clock, LOW);
    digitalWrite(clock, HIGH);
  }
  // Turn the levels off
  digitalWrite(Level1, LOW);
  digitalWrite(Level2, LOW);
  digitalWrite(Level3, LOW);
  digitalWrite(Level4, LOW);
  digitalWrite(Level5, LOW);
  digitalWrite(Level6, LOW);
  digitalWrite(Level7, LOW);
  digitalWrite(Level8, LOW);
  //Latches in the new data
  digitalWrite(latch, LOW);
  digitalWrite(latch, HIGH);
  // Select the new level to turn on
  if (level==0)digitalWrite(Level1, HIGH);
  if (level==1)digitalWrite(Level2, HIGH);
  if (level==2)digitalWrite(Level3, HIGH);
  if (level==3)digitalWrite(Level4, HIGH);
  if (level==4)digitalWrite(Level5, HIGH);
  if (level==5)digitalWrite(Level6, HIGH);
  if (level==6)digitalWrite(Level7, HIGH);
  if (level==7)digitalWrite(Level8, HIGH);
  delay(1);
  }
}
```

Here is another demo program, this one takes the random number and shifts it down producing a "falling rain" kind of effect.

```
// 8x8x8 LED Cube with 74595's falling rain demo

// These Pins Connect to 74595's
int data = 2;
int clock = 4;
int latch = 3;
// These Pins Connect to ULN2803
```

```
int Level1 = 5;
int Level2 = 6;
int Level3 = 7;
int Level4 = 8;
int Level5 = 9;
int Level6 = 10;
int Level7 = 11;
int Level8 = 12;

int rotate = 0;
int cycle = 0;
// stock up on random numbers
int rdata1 = random(64);
int rdata2 = random(64);
int rdata3 = random(64);
int rdata4 = random(64);
int rdata5 = random(64);
int rdata6 = random(64);
int rdata7 = random(64);
int rdata8 = random(64);

// set up output pins
void setup() {
  pinMode(data, OUTPUT);
  pinMode(clock, OUTPUT);
  pinMode(latch, OUTPUT);
  pinMode(Level1, OUTPUT);
  pinMode(Level2, OUTPUT);
  pinMode(Level3, OUTPUT);
  pinMode(Level4, OUTPUT);
  pinMode(Level5, OUTPUT);
  pinMode(Level6, OUTPUT);
  pinMode(Level7, OUTPUT);
  pinMode(Level8, OUTPUT);
}

void loop() {
  // stock up on random numbers.
  for (int level=0; level <8; level++){
    for (int shift=0; shift <64; shift++){
      digitalWrite(data, LOW);
```

```
  // select the random number for the correct level
  if (level==0 and rdata1==shift) digitalWrite(data, HIGH);
  if (level==1 and rdata2==shift) digitalWrite(data, HIGH);
  if (level==2 and rdata3==shift) digitalWrite(data, HIGH);
  if (level==3 and rdata4==shift) digitalWrite(data, HIGH);
  if (level==4 and rdata5==shift) digitalWrite(data, HIGH);
  if (level==5 and rdata6==shift) digitalWrite(data, HIGH);
  if (level==6 and rdata7==shift) digitalWrite(data, HIGH);
  if (level==7 and rdata8==shift) digitalWrite(data, HIGH);
  // Clocks in the new data
  digitalWrite(clock, LOW);
  digitalWrite(clock, HIGH);
}
// Turn the levels off
digitalWrite(Level1, LOW);
digitalWrite(Level2, LOW);
digitalWrite(Level3, LOW);
digitalWrite(Level4, LOW);
digitalWrite(Level5, LOW);
digitalWrite(Level6, LOW);
digitalWrite(Level7, LOW);
digitalWrite(Level8, LOW);
 //Latches in the new data
digitalWrite(latch, LOW);
digitalWrite(latch, HIGH);
// Select the new level to turn on
if (level==0)digitalWrite(Level1, HIGH);
if (level==1)digitalWrite(Level2, HIGH);
if (level==2)digitalWrite(Level3, HIGH);
if (level==3)digitalWrite(Level4, HIGH);
if (level==4)digitalWrite(Level5, HIGH);
if (level==5)digitalWrite(Level6, HIGH);
if (level==6)digitalWrite(Level7, HIGH);
if (level==7)digitalWrite(Level8, HIGH);
delay(1);
}
// Rate of movement - one per 8 cycles
rotate = rotate+1;
if (rotate > 7) {
 rotate=0;
 // shift the random numbers down
```

```
    rdata8 = rdata7;
    rdata7 = rdata6;
    rdata6 = rdata5;
    rdata5 = rdata4;
    rdata4 = rdata3;
    rdata3 = rdata2;
    rdata2 = rdata1;
    rdata1 = random(64);
  }
}
```

Chapter 7

8x8x8 LED Cube kit

The LED arrays and cross connects in the previous project are completely compatible with this 8x8x8 LED circuit board kit that is pictured below. The kit uses latches instead of shift registers so there is no easy way to interface the Arduino into the kit. But the Arduino retrofit can be done. The Arduino 8x8x8 cube retrofit is a project by itself. It shares the eight LED arrays and the circuit board shown below. Almost all of the IC's will need to be replaced.

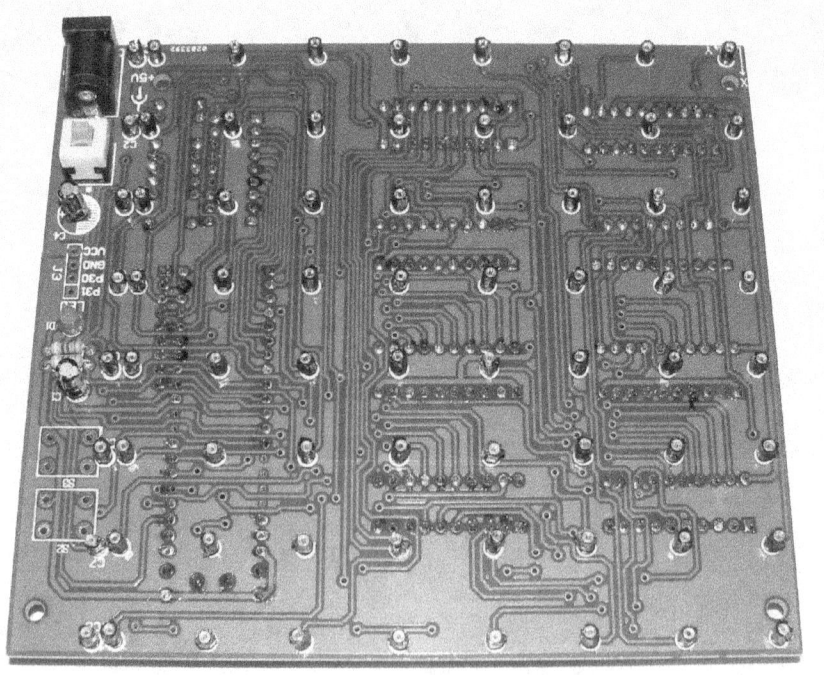

Here is a picture of the other side of the 8x8x8 LED kit control board. If you change the eight resistors in the upper right corner to 50 ohms instead of 500 ohms you will get a much brighter display.

The next picture shows the completed 8x8x8 LED cube kit.

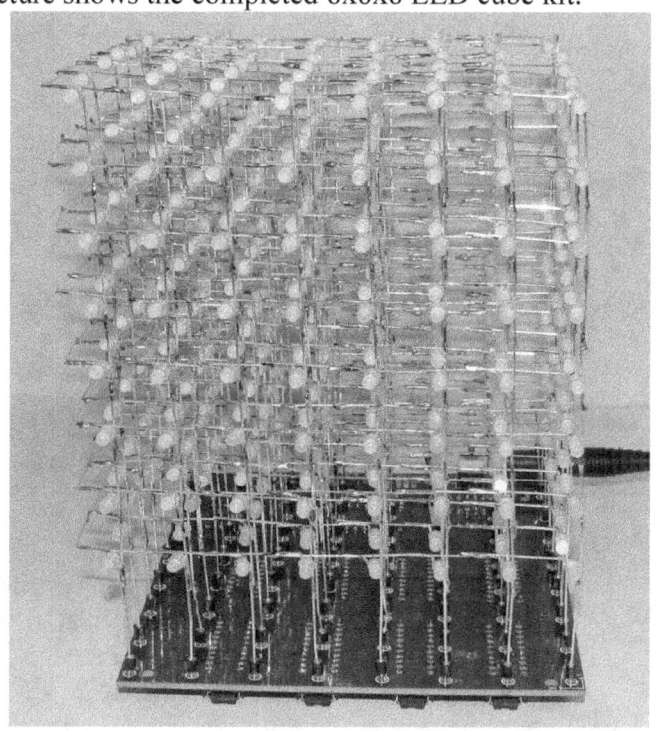

Here is a back view of the 8x8x8 LED cube kit.

Now for the 8x8x8 LED cube kit retrofit project. It involves removing the CPU and adding an adapter that connects to the Arduino. First I had to meter out the pins of the current CPU chip. This picture shows the important pins.

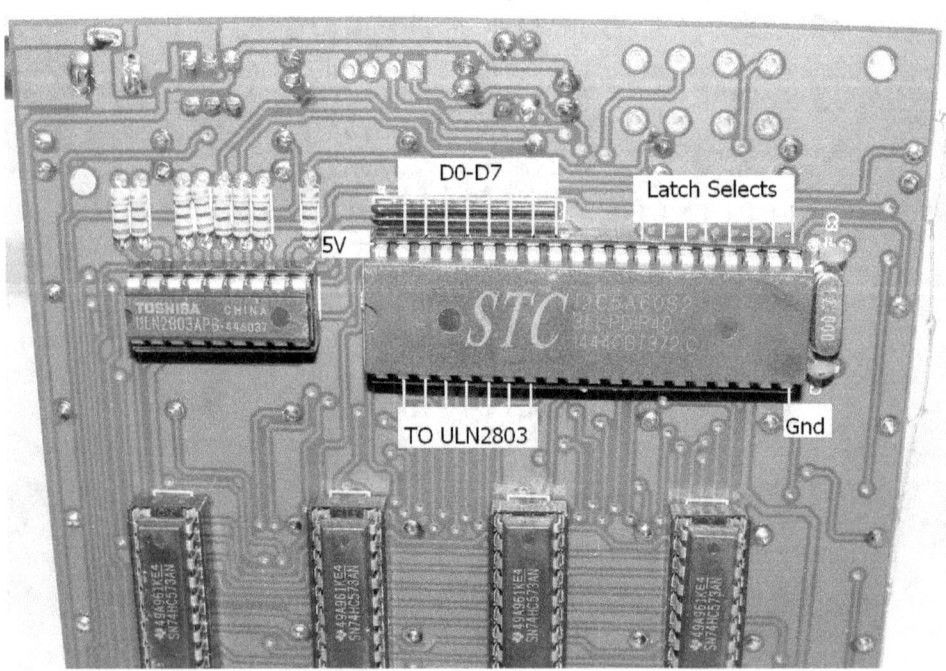

Then you will need to make a 40 pin adapter that plugs into the old CPU socket and has the pins to interface to the Arduino. Here is the schematic diagram for the adapter.

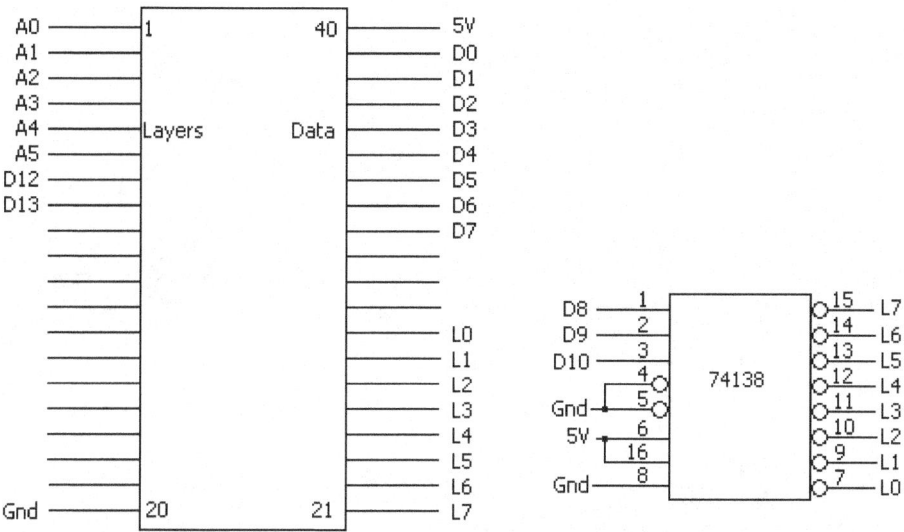

This is what the adapter looks like when it is wired up. Make sure you change the eight latch IC's to 74HC574's. Otherwise all the latches and LED's will display the same thing because the old latches are "transparent" latches.

This is the resulting schematic diagram that crosses the adapter with the existing circuit board and parts.

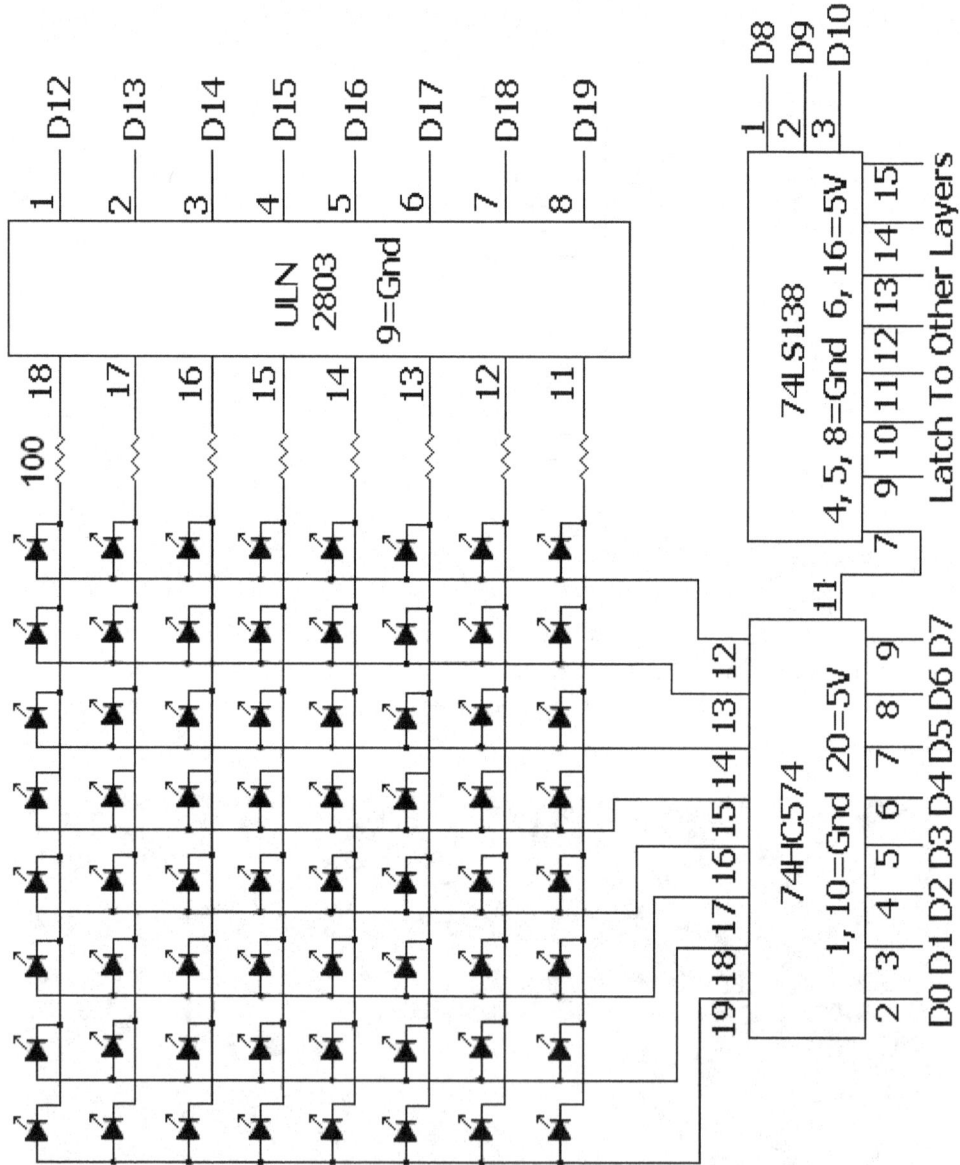

Up next is a picture of the completed Arduino retrofit. I tested it with some programs that I found on the Internet for compatibility with their design. It works with most Arduino 8x8x8 LED cube designs that use eight 74HC574 latches and a 74138 latch selection chip.

I am only listing two demo programs for this cube because there are so many nice demo programs available on the Internet. But none of the other demo programs do my diagonal slice so here it is.

```
// 8x8x8 kit retrofitted with arduino slice demo
// uses some ports in parallel mode
// PORTD = D0 to D7 Data port to latches
// PORTB = D8, D9, D10 or bits 0, 1 and 2 are latch select
// PORTB = D12, D13 or bits 4 and 5 are level select 6 and 7
// PORTC = A0 to A5 are level select 0-5
int cycle = 0;
int level = 0;
int shift = 0;
int latch = 0;
int i;
byte cube[][8]= {
// there are eight bytes of eight LED's per layer.
 {0x80, 0x40, 0x20, 0x10, 0x08, 0x04, 0x02, 0x01},
 {0x40, 0x20, 0x10, 0x08, 0x04, 0x02, 0x01, 0x80},
```

```
{0x20, 0x10, 0x08, 0x04, 0x02, 0x01, 0x80, 0x40},
{0x10, 0x08, 0x04, 0x02, 0x01, 0x80, 0x40, 0x20},
{0x08, 0x04, 0x02, 0x01, 0x80, 0x40, 0x20, 0x10},
{0x04, 0x02, 0x01, 0x80, 0x40, 0x20, 0x10, 0x08},
{0x02, 0x01, 0x80, 0x40, 0x20, 0x10, 0x08, 0x04},
{0x01, 0x80, 0x40, 0x20, 0x10, 0x08, 0x04, 0x02},
{0x80, 0x40, 0x20, 0x10, 0x08, 0x04, 0x02, 0x01},
{0x40, 0x20, 0x10, 0x08, 0x04, 0x02, 0x01, 0x80},
{0x20, 0x10, 0x08, 0x04, 0x02, 0x01, 0x80, 0x40},
{0x10, 0x08, 0x04, 0x02, 0x01, 0x80, 0x40, 0x20},
{0x08, 0x04, 0x02, 0x01, 0x80, 0x40, 0x20, 0x10},
{0x04, 0x02, 0x01, 0x80, 0x40, 0x20, 0x10, 0x08},
{0x02, 0x01, 0x80, 0x40, 0x20, 0x10, 0x08, 0x04},
{0x01, 0x80, 0x40, 0x20, 0x10, 0x08, 0x04, 0x02},
};

// set all pins to output mode
void setup(){
  for(i=0; i<20; i++) pinMode(i, OUTPUT);
  }

void loop(){
  for (level=0; level<8; level++) {
    // turn all level selects off
    PORTC = 0x00;
    digitalWrite(12, LOW);
    digitalWrite(13, LOW);
    // load latches for current layer
    for (latch=0; latch<8; latch++) {
      // load data
      PORTD = cube[level+shift][latch];
      // PORTD = 0xFF; // all on test
      // Select next latch
      PORTB = (latch);
    }

  // Select the level and turn it on
  if (level == 0) digitalWrite(14, HIGH);
  if (level == 1) digitalWrite(15, HIGH);
  if (level == 2) digitalWrite(16, HIGH);
  if (level == 3) digitalWrite(17, HIGH);
```

```
  if (level == 4) digitalWrite(18, HIGH);
  if (level == 5) digitalWrite(11, HIGH);
  if (level == 6) digitalWrite(12, HIGH);
  if (level == 7) {
    digitalWrite(13, HIGH);
    cycle = cycle+1;
    if (cycle > 7) {
      cycle = 0;
      shift = shift+1;
      if (shift > 7) shift = 0;
    }
  }
  delay(2);
  }
}
```

This second demo program allows the 8x8x8 LED cube kit to display sound. You might think that all of the Arduino pins are in use. However it turns out that the cube enable pin D11 is not needed, it can be moved to a ground pin. Then the level select pin on A5 can be moved to D11 so you now have a spare analog input pin. Here is the code to display analog or sound on the cube.

```
// 8x8x8 kit retrofitted with arduino sound demo
// This program uses some ports in parallel mode
// PORTD = D0 to D7 Data port to latches
// PORTB = D8, D9, D10 or bits 0, 1, 2 are latch select
// PORTB = D11, D12, D13 or bits 3, 4, 5 are level select 5, 6, 7
// PORTC = A0 to A4 are level selects 0-4 as D14 to D18
// PORTC = A5 is analog input <- NOTE THIS CHANGE
int level = 0;
int latch = 0;
int i;
// values for the X axis
byte cube[8]= {0x80, 0xc0, 0xe0, 0xf0, 0xf8, 0xfc, 0xfe, 0xff};

// set all pins to output mode
void setup(){
  // Leave A5 as an analog input
  for(i=0; i<19; i++) pinMode(i, OUTPUT);
  }
```

```
void loop(){
  for (level=0; level<8; level++) {
    // turn all level selects off
    PORTC = 0x00;
    PORTB = 0x00;
    // Get analog sample and reduce it to 0-64
    int analog = analogRead(5)/8;
    // Get X and Y axis position
    int yaxis = analog/8;
    int xaxis = analog%8;
    for (latch=0; latch<8; latch++) {
      // load data
      if (latch > yaxis+1) PORTD = 0x00;
      if (latch < yaxis+1) PORTD = 0xFF;
      if (latch == yaxis+1) PORTD = cube[xaxis];
      // PORTD = 0xFF; // all LED's on test
      // Select latch
      PORTB = (latch);
    }

    // Select the level and turn it on
    if (level == 0) digitalWrite(14, HIGH);
    if (level == 1) digitalWrite(15, HIGH);
    if (level == 2) digitalWrite(16, HIGH);
    if (level == 3) digitalWrite(17, HIGH);
    if (level == 4) digitalWrite(18, HIGH);
    if (level == 5) digitalWrite(11, HIGH);
    if (level == 6) digitalWrite(12, HIGH);
    if (level == 7) digitalWrite(13, HIGH);
    delay(2);
  }
}
```

Chapter 8

4x4x4 Color LED Cube

Someone said that a common anode setup was brighter. I accidentally ordered some RGB LED's that were common anode so I though I would give it a try. Also this book would not be complete without a 4x4x4 RGB cube so this chapter will kill two birds with one stone. Common anode requires a driver IC that can deliver 5V instead of ground like the ULN2803 does. The easiest way to get a 5V driver is to make one, so I made one out of TIP125's .

The grid size for this project is .8 inches and that is just barely enough space for the LED leads to reach to the next LED. To be honest I sometimes had to bridge a small gap with the solder. The holes are 3/16 inch in diameter for the 5mm LED's. The holes could be just a tiny bit larger as that hole size made it hard to remove the array once the array was completed.

To get started, first mark off the .8 by .8 inch grid on a piece of wood about 4 inches by four inches in size. You can also use 2.0 cm by 2.0 cm as most rulers are not marked in tenths of an inch. Then drill the 16 holes with a 3/16 inch drill bit. If you have a drill bit that is a tiny bit larger then use that, otherwise run the drill bit in and out to make the hole slightly larger. You should have a resulting hole size that allows the LED to fit snugly but still be easy to remove.

It is best to mark the board showing what way the common and the RGB leads are to go so the arrays will all end up being wired the same way.

On the RGB LED the common lead is the longest lead. I always oriented the longer lead towards the right. Then bend the two outside leads down at a 45 degree angle. Next bend the outside leads straight down at a second 45 degree angle about 1/8 inch form the LED.

Up next are pictures of a typical RGB LED and a picture of the outside pins being bent on the two 45 degree angles.

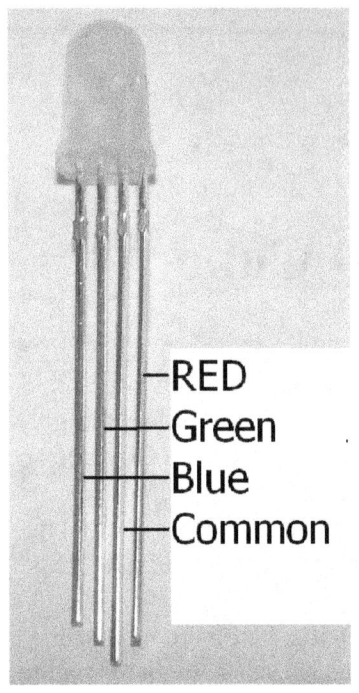
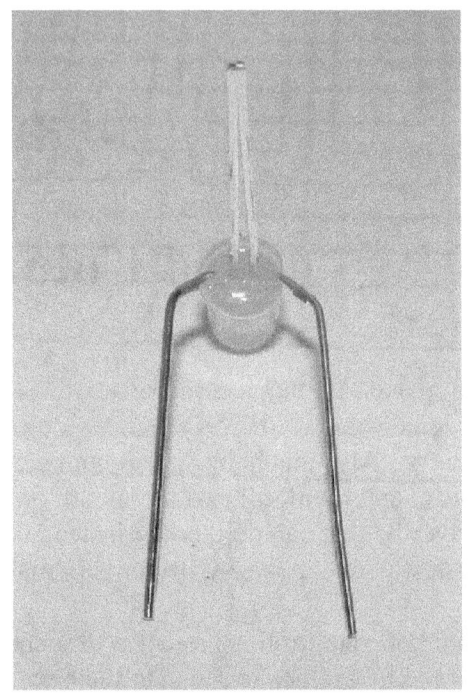

The arrays are made the same way that they are made for the 8x8x8 color LED cube. First the outside leads are bent on two 45 degree angles.

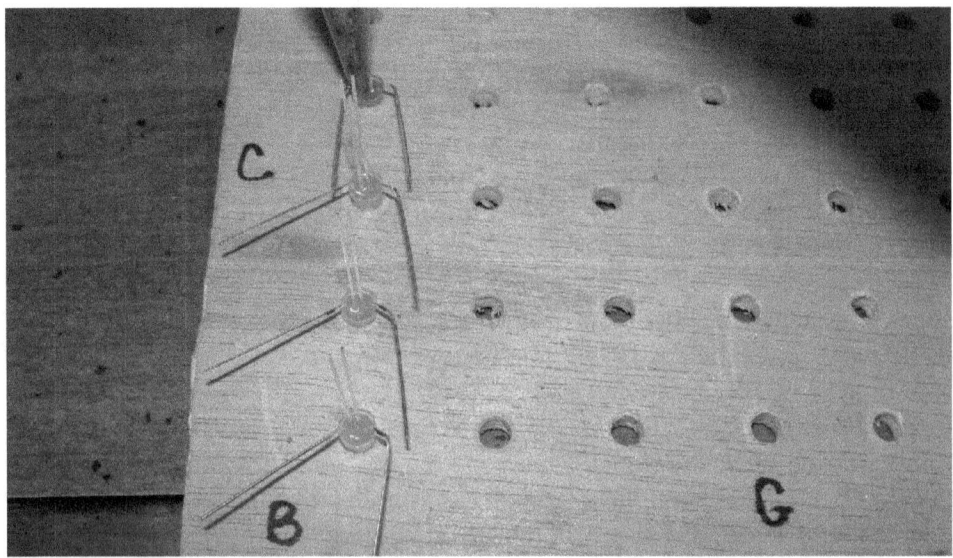

Once they are aligned so they are close together they are soldered together.

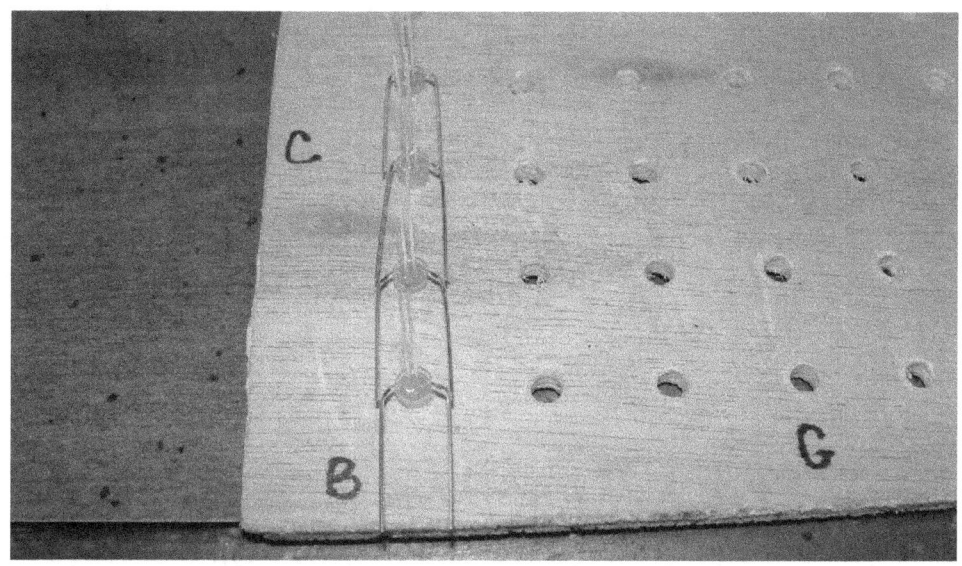

Then the shorter of the two remaining leads is bent down and soldered together. Then the longest lead is bent about 1/4 inches from the LED at a 90 degree angle. A pair of needle nose pliers can make sure they are all about the same length. They then are soldered together.

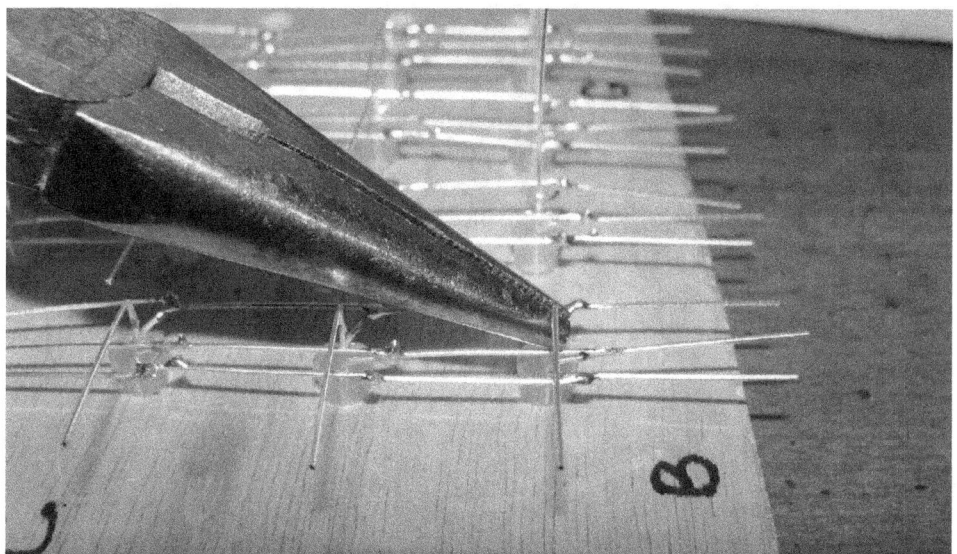

Next the current limiting resistors are added, completing the arrays.

Up next is a picture of what the completed project looks like.

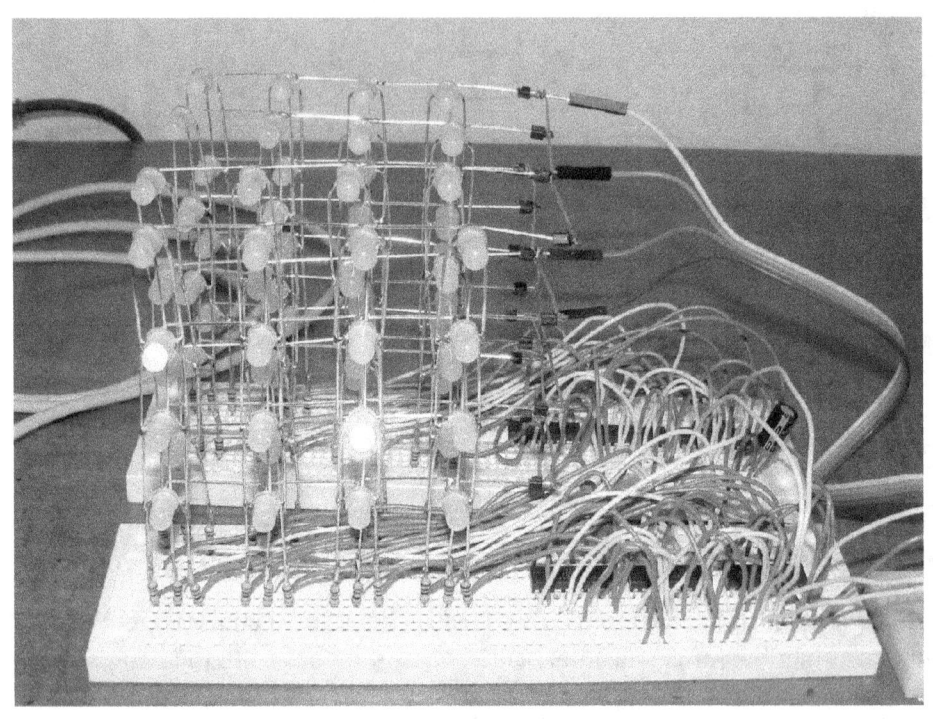

Here is a picture of the cube from above.

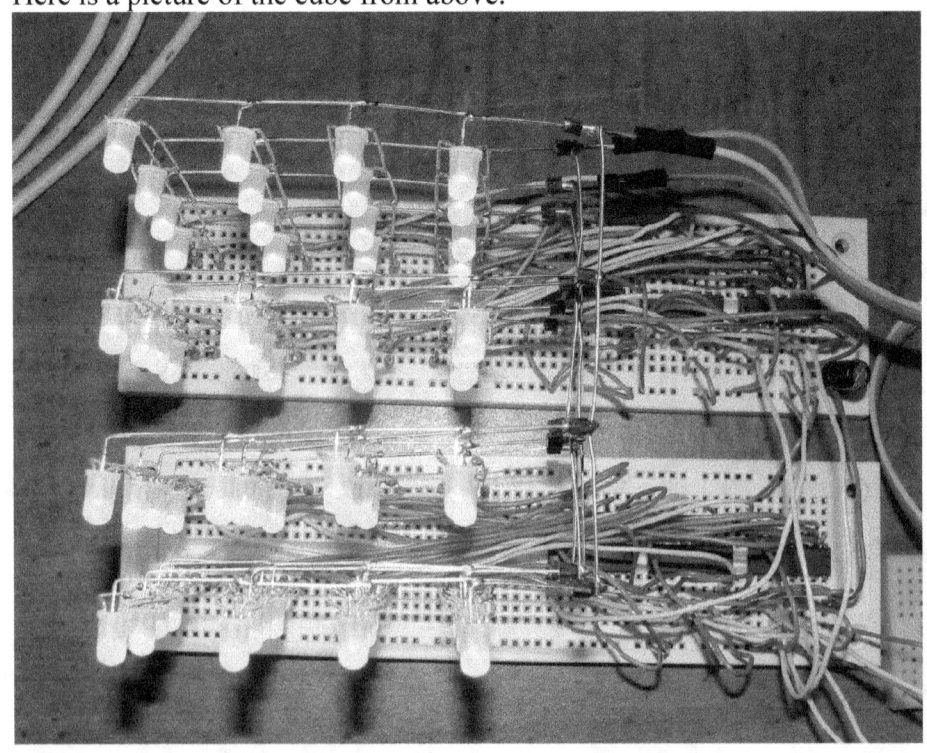

This is the schematic diagram of the 4x4x4 RGB Cube. This should fit on one breadboard. Two of these are made for the complete project.

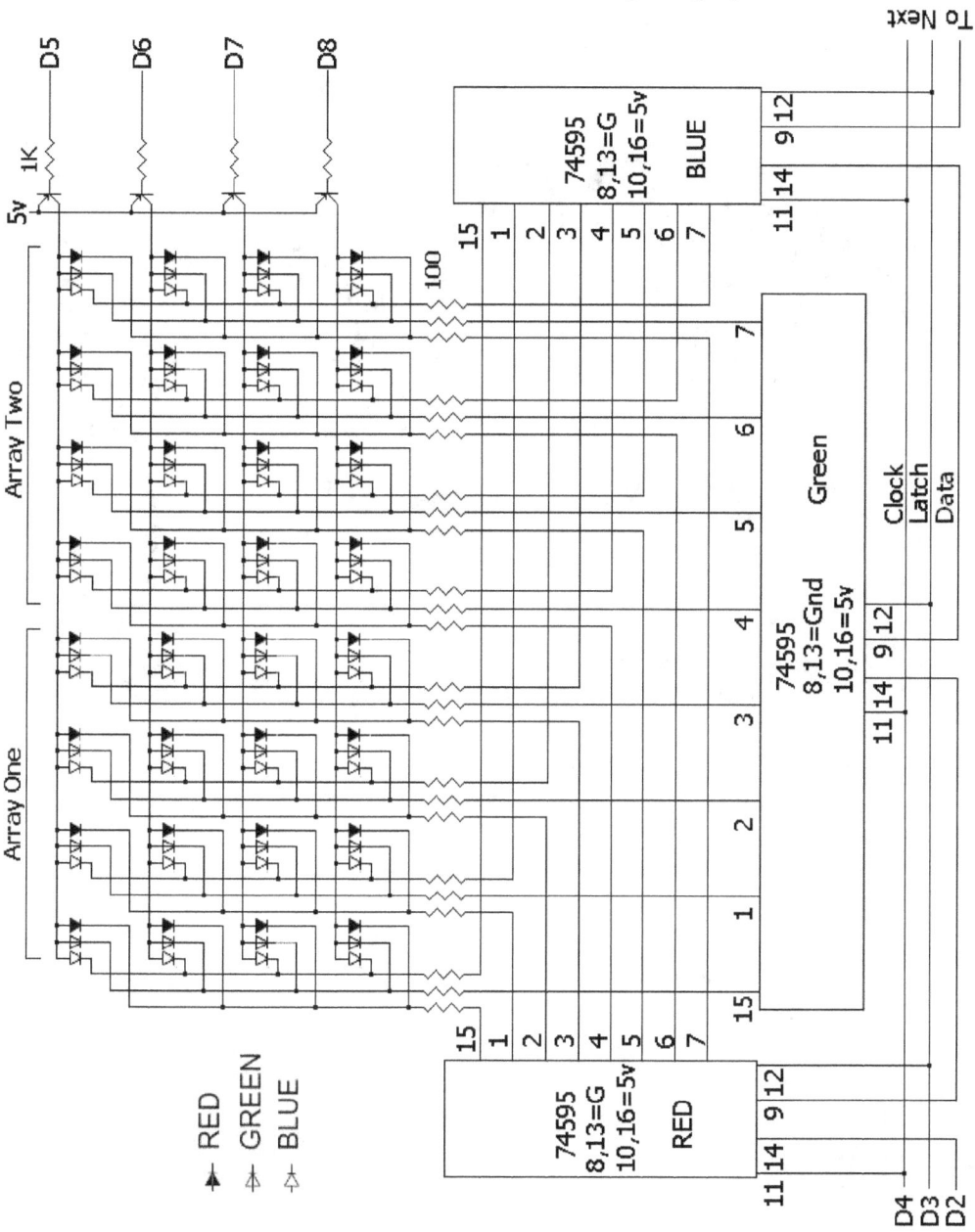

Now for a word or two to explain why the TIP125 or TIP127 common anode driver transistors are mounted on a DIP header. If you plug them into the breadboard their leads may damage the breadboard. Their leads are a little too wide for the breadboard contacts and may permanently bend them. The solution is to solder them into a dip header and then plug that in like an IC. I should have included the 1K or 2K resistors in the header but did not. Here is a close up picture of the driver transistors. The base (input) is on the left, the collector is in the middle and the emitter is on the right.

Here are three demo programs. Remember everything is inverted. A high=off and a low=on. This first demo tests all of the LED's to see if they are working.

```
// 4x4x4 RGB Cube testing all LED's

// These Pins Connect to 74595's
int data = 2;
int latch = 3;
int clock = 4;
// These Pins Connect to TIP125's
int Level1 = 5;
int Level2 = 6;
int Level3 = 7;
int Level4 = 8;

int rotate = 0;
int cycle = 0;
```

```
// set up output pins
void setup() {
  pinMode(data, OUTPUT);
  pinMode(clock, OUTPUT);
  pinMode(latch, OUTPUT);
  pinMode(Level1, OUTPUT);
  pinMode(Level2, OUTPUT);
  pinMode(Level3, OUTPUT);
  pinMode(Level4, OUTPUT);
}

void loop() {
  for (int level=0; level <4; level++){
    for (int shift=0; shift <48; shift++){
      // Remember data is inverted!
      digitalWrite(data, HIGH);
      if (rotate == 0){
        if (shift >= 0 and shift < 8){digitalWrite(data, LOW);}
        if (shift > 23 and shift < 32){digitalWrite(data, LOW);}
      }
      if (rotate == 1){
        if (shift > 7 and shift < 16){digitalWrite(data, LOW);}
        if (shift > 31 and shift < 40){digitalWrite(data, LOW);}
      }
      if (rotate == 2){
        if (shift > 15 and shift < 24){digitalWrite(data, LOW);}
        if (shift > 39 and shift < 48){digitalWrite(data, LOW);}
      }
      if (rotate == 3){digitalWrite(data, LOW);}
      // Clocks in the new data
      digitalWrite(clock, LOW);
      digitalWrite(clock, HIGH);
    }
    // Turn the levels off
    // Remember its inverted for common anode!
    digitalWrite(Level1, HIGH);
    digitalWrite(Level2, HIGH);
    digitalWrite(Level3, HIGH);
    digitalWrite(Level4, HIGH);
    //Latch in the new data
```

```
    digitalWrite(latch, LOW);
    digitalWrite(latch, HIGH);
    // Select the new level to turn on
    if (level==0)digitalWrite(Level1, LOW);
    if (level==1)digitalWrite(Level2, LOW);
    if (level==2)digitalWrite(Level3, LOW);
    if (level==3)digitalWrite(Level4, LOW);
    delay(2);
  }
  // Rate of movement - one per 50 cycles
  cycle=cycle+1;
  if (cycle == 50) {
    rotate = rotate+1;
    cycle = 0;
  }
  if (rotate > 4) rotate=0;
}
```

This next demo program uses a random number to light random LED's.
They flash kind of like fireworks.

```
// 4x4x4 RGB Cube with Random sparkles

// These Pins Connect to 74595's
int data = 2;
int latch = 3;
int clock = 4;
// These Pins Connect to ULN2003
int Level1 = 5;
int Level2 = 6;
int Level3 = 7;
int Level4 = 8;

int rotate = 0;
int cycle = 0;

// set up output pins
void setup() {
  pinMode(data, OUTPUT);
  pinMode(clock, OUTPUT);
  pinMode(latch, OUTPUT);
```

```
  pinMode(Level1, OUTPUT);
  pinMode(Level2, OUTPUT);
  pinMode(Level3, OUTPUT);
  pinMode(Level4, OUTPUT);
}

void loop() {
  for (int level=0; level <4; level++){
   int rdata = random(48);
   for (int shift=0; shift <48; shift++){
    // Data is inverted!
    digitalWrite(data, HIGH);
    if (rdata==shift){
     digitalWrite(data, LOW);
    }
    // Clocks in the new data
    digitalWrite(clock, LOW);
    digitalWrite(clock, HIGH);
   }
   // Turn the levels off
   digitalWrite(Level1, HIGH);
   digitalWrite(Level2, HIGH);
   digitalWrite(Level3, HIGH);
   digitalWrite(Level4, HIGH);
    //Latches in the new data
   digitalWrite(latch, LOW);
   digitalWrite(latch, HIGH);
   // Select the new level to turn on
   if (level==0)digitalWrite(Level1, LOW);
   if (level==1)digitalWrite(Level2, LOW);
   if (level==2)digitalWrite(Level3, LOW);
   if (level==3)digitalWrite(Level4, LOW);
   delay(50);
  }
}
```

This last demo program is the falling rain demo but this time it is in color.

```
// 4x4x4 RGB Cube with falling rain

// These Pins Connect to 74595's
```

```
int data = 2;
int latch = 3;
int clock = 4;
// These Pins Connect to ULN2803
int Level1 = 5;
int Level2 = 6;
int Level3 = 7;
int Level4 = 8;

int rotate = 0;
int cycle = 0;
// stock up on random numbers
int rdata1 = random(96);
int rdata2 = random(96);
int rdata3 = random(96);
int rdata4 = random(96);

// set up output pins
void setup() {
  pinMode(data, OUTPUT);
  pinMode(clock, OUTPUT);
  pinMode(latch, OUTPUT);
  pinMode(Level1, OUTPUT);
  pinMode(Level2, OUTPUT);
  pinMode(Level3, OUTPUT);
  pinMode(Level4, OUTPUT);
}

void loop() {
  // stock up on random numbers.
  for (int level=0; level <4; level++){
    for (int shift=0; shift <48; shift++){
      // data is inverted!
      digitalWrite(data, HIGH);
      // select the random number for the correct level
      if (level==0 and rdata1==shift) digitalWrite(data, LOW);
      if (level==1 and rdata2==shift) digitalWrite(data, LOW);
      if (level==2 and rdata3==shift) digitalWrite(data, LOW);
      if (level==3 and rdata4==shift) digitalWrite(data, LOW);
      digitalWrite(clock, LOW);
      digitalWrite(clock, HIGH);
```

```
  }
  // Turn the levels off
  digitalWrite(Level1, HIGH);
  digitalWrite(Level2, HIGH);
  digitalWrite(Level3, HIGH);
  digitalWrite(Level4, HIGH);
   //Latches in the new data
  digitalWrite(latch, LOW);
  digitalWrite(latch, HIGH);
  // Select the new level to turn on
  if (level==0)digitalWrite(Level1, LOW);
  if (level==1)digitalWrite(Level2, LOW);
  if (level==2)digitalWrite(Level3, LOW);
  if (level==3)digitalWrite(Level4, LOW);
  delay(2);
}
// Rate of movement - one per 8 cycles
rotate = rotate+1;
if (rotate > 12) {
  rotate=0;
  // shift the random numbers down
  rdata4 = rdata3;
  rdata3 = rdata2;
  rdata2 = rdata1;
  rdata1 = random(48);
 }
}
```

Chapter 9

8x8x8 Color LED Cube

This last project is by far the most complex one. The design supports an 8x8x8 Color LED cube, but I only made a 7x7x4 Color cube initially because I only had 200 common cathode RGB LED's. Also, it is not easy to wire up the middle layers of the LED arrays to make it more than 4 layers deep.

The grid size for this project is .8 inches and that is the same as the 4x4x4 RGB cube. You might have to bridge some small gaps with the solder. The holes are 3/16 inch in diameter for the 5mm LED's. The holes could be just a tiny bit larger as using that size made it hard to remove the array once the array was completed.

To get started, first mark off the .8 by .8 inch grid on a piece of wood about 8 inches by 8 inches in size. You can also use 2.0 cm by 2.0 cm as most rulers are not marked in tenths of an inch. Then drill the 64 holes with a 3/16 inch drill bit. If you have a drill bit that is a tiny bit larger then use it, otherwise run the drill bit in and out or use a file to make the hole slightly larger. You should have a hole that allows the LED to fit snugly but still be easy to remove.

It is best to mark the board showing what way the common and the RGB leads are to go so the arrays will all end up being wired the same way. I have actually ended up with one array that was different than all of the others on each of these LED cube projects!

Like on the 4x4x4 color cube the common lead is the longest lead. Orient the longer lead towards the right side of the board. Then bend the two outside leads down at a 45 degree angle. Next bend the outside leads straight down at a second 45 degree angle about 1/8 inch form the LED.

Then this next picture shows how the LED's should align once they are inserted into the board.

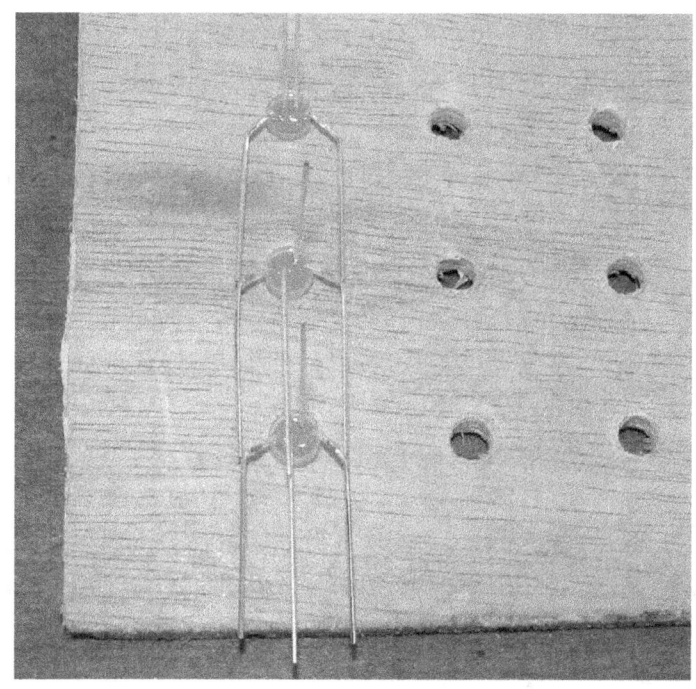

Now you can solder the outside leads sequentially together down each row.

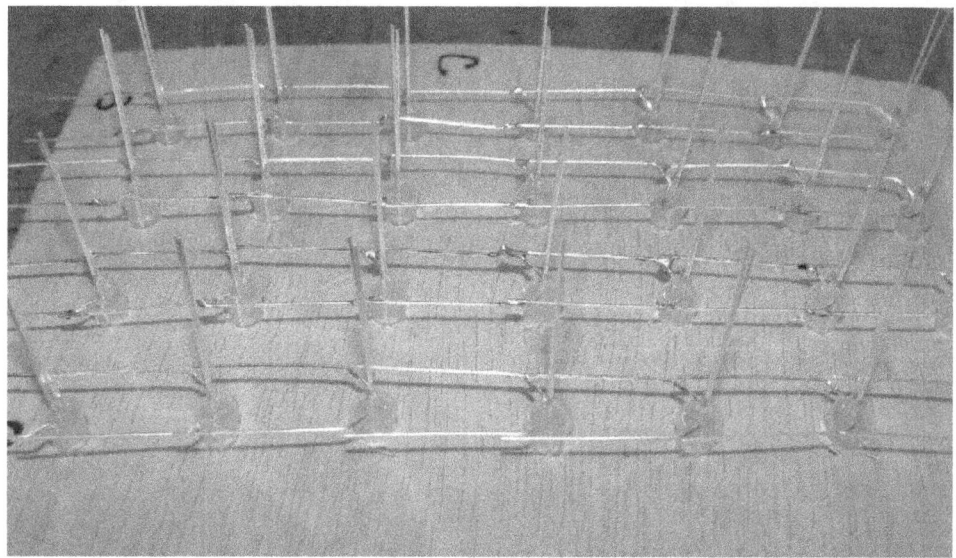

Once the outside leads are soldered in place you can then solder the shorter of the two remaining leads together. They bend straight down from one LED to the next LED. Be sure to solder them with the iron at least 1/4 an inch down from the LED itself. Do not apply heat directly on the LED or it will melt.

The last and longest lead is the common. It is bent in the other or perpendicular direction around 1/8 of an inch out from the LED. Then they are soldered together as well.

The next picture shows what one of the color arrays looked like when it was half way done. I bent the leads once the LED was already inserted in the hole but it might be better to bend them ahead of time in some sort of a bending jig to get them to all be the same.

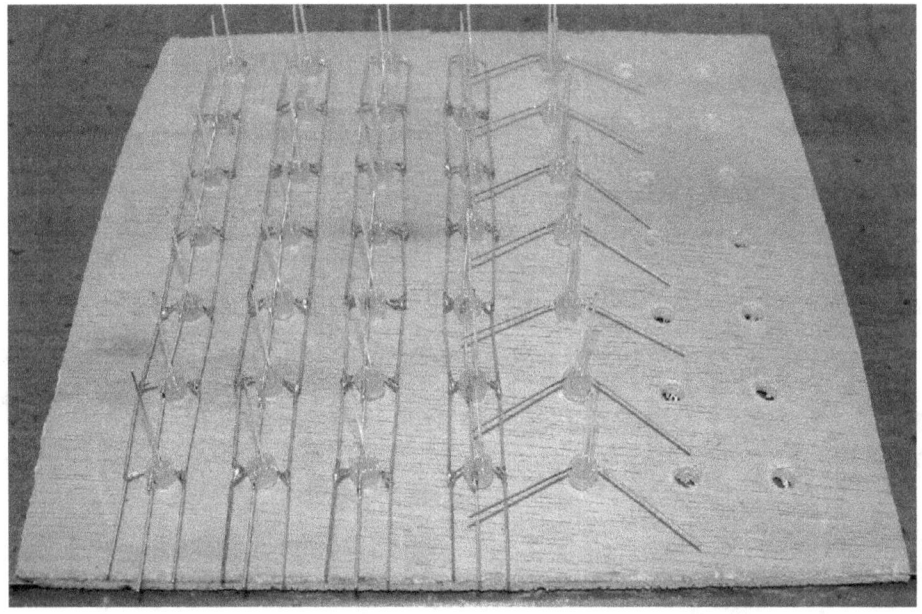

The next picture shows a close up of the cross connected longest, common cathode, lead. I use a pair of needle nose pliers to position the bend so they are all about the same length.

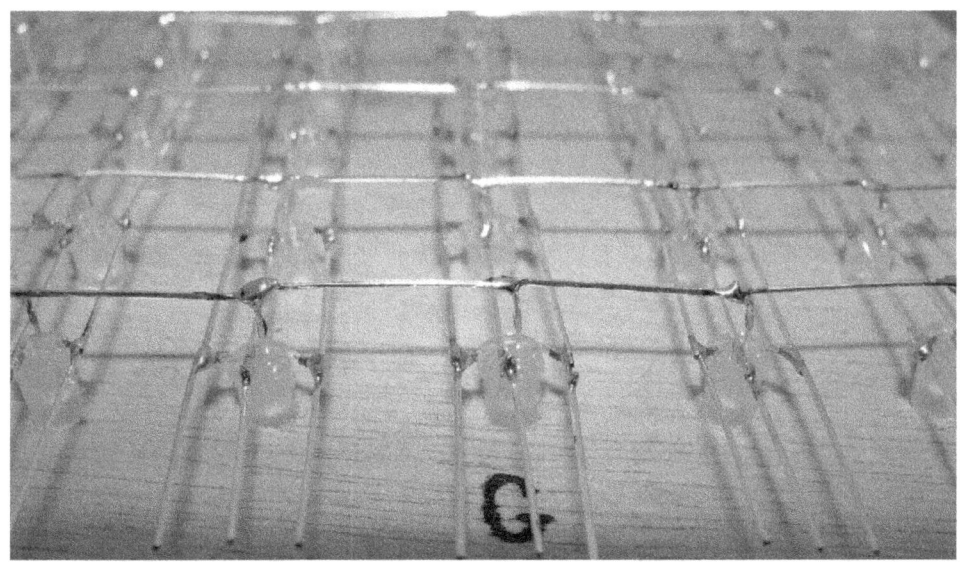

The next picture shows an almost completed color 7x7 array, the last step is to add the current limiting resistors.

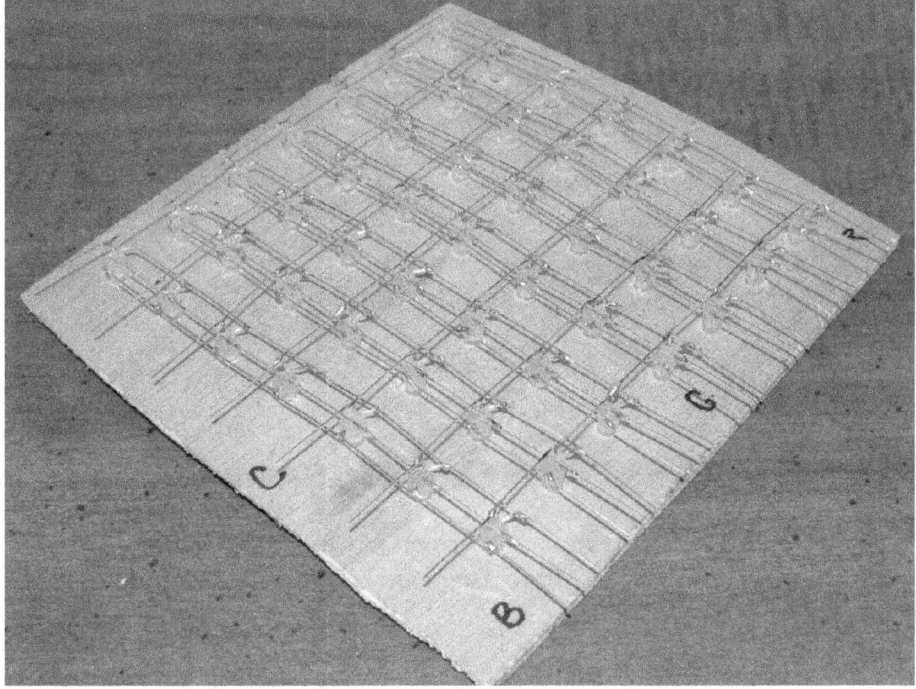

One last step is to add the resistors. I did that in the breadboard. First I inserted the resistors into every other position. Then remove every fourth resistor. This is the pattern of resistors that you should end up with.

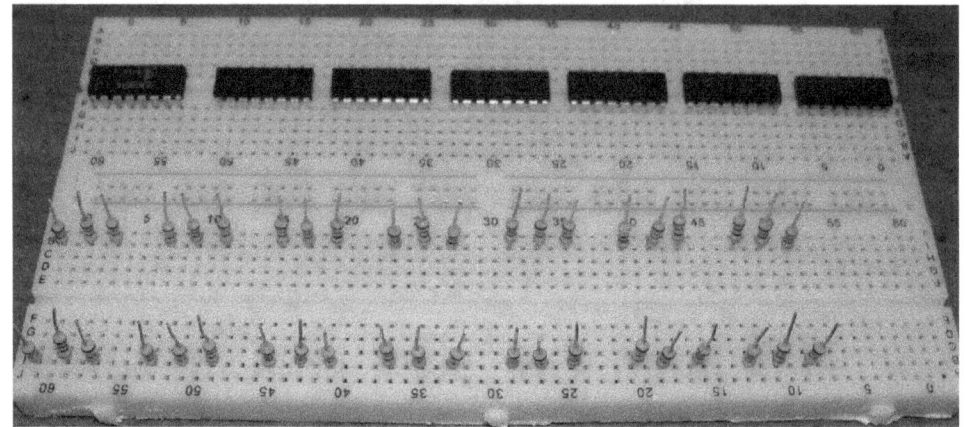

Next align the array with the resistors and solder the two outside LED's to their resistors. Then using needle nose pliers to align them, solder the other LED's to their resistors. This picture is what they should look like.

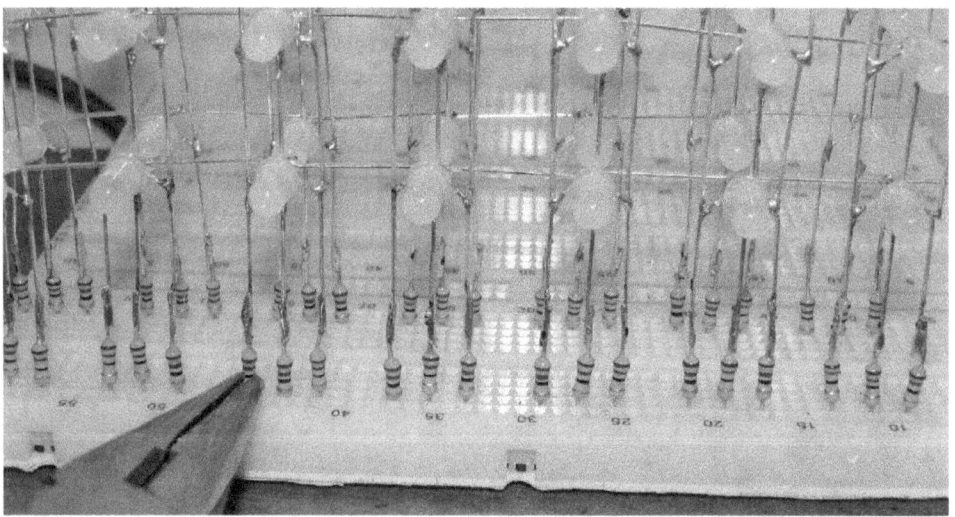

This next picture shows the front view when just the front two arrays were wired up and working. The shift registers are oriented so their outputs face the LED arrays.

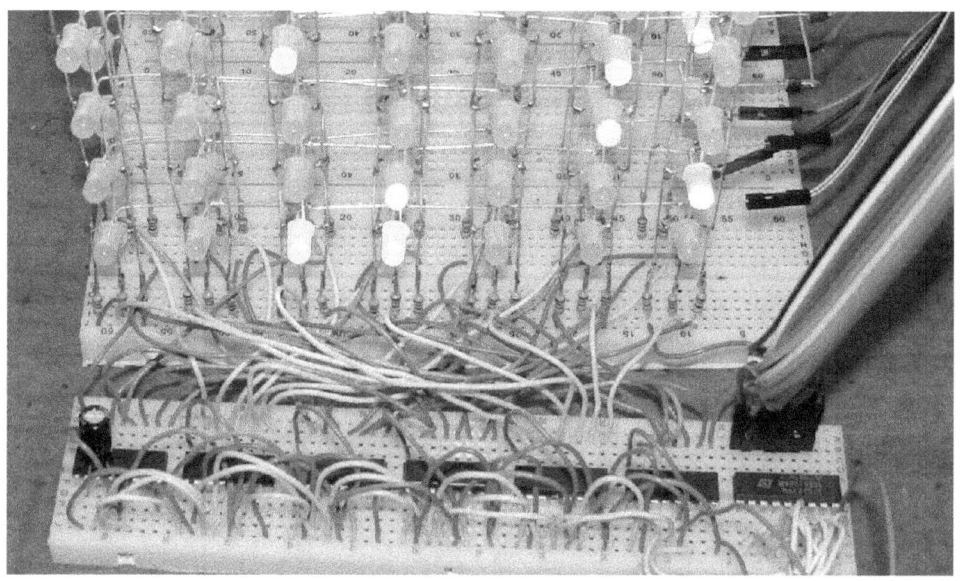

This is a view of the back side showing the shift registers. It is wired as a mirrored arrangement of the front. The IC's are reversed so their outputs again face the LED arrays. Long wires, going around the right side, carry power, clock, latch and data from the front to the back.

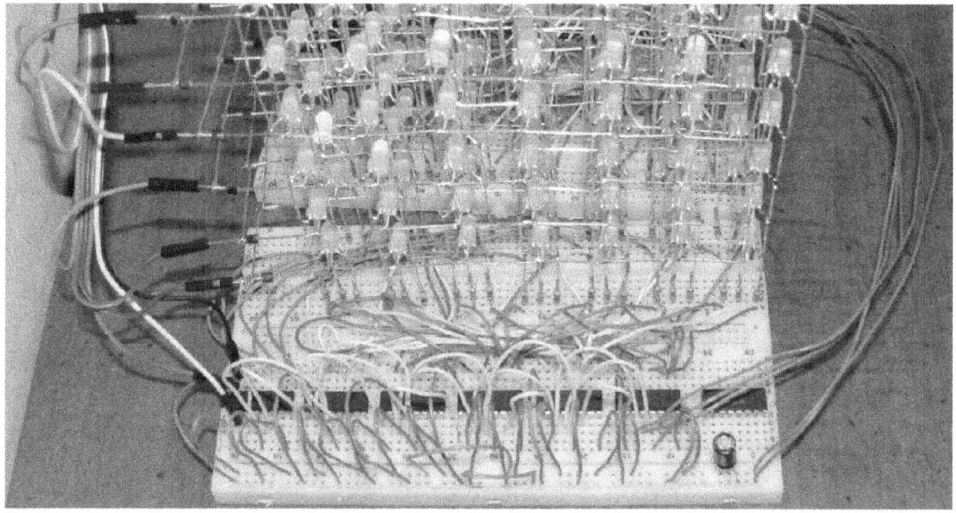

Here is the schematic diagram for the 8x8x8 color cube. It is so big that it had to be split between two pages. The ULN2803 can handle four of the 8x8 arrays but you might need to add a second ULN2803 if you are adding four more 8x8 arrays.

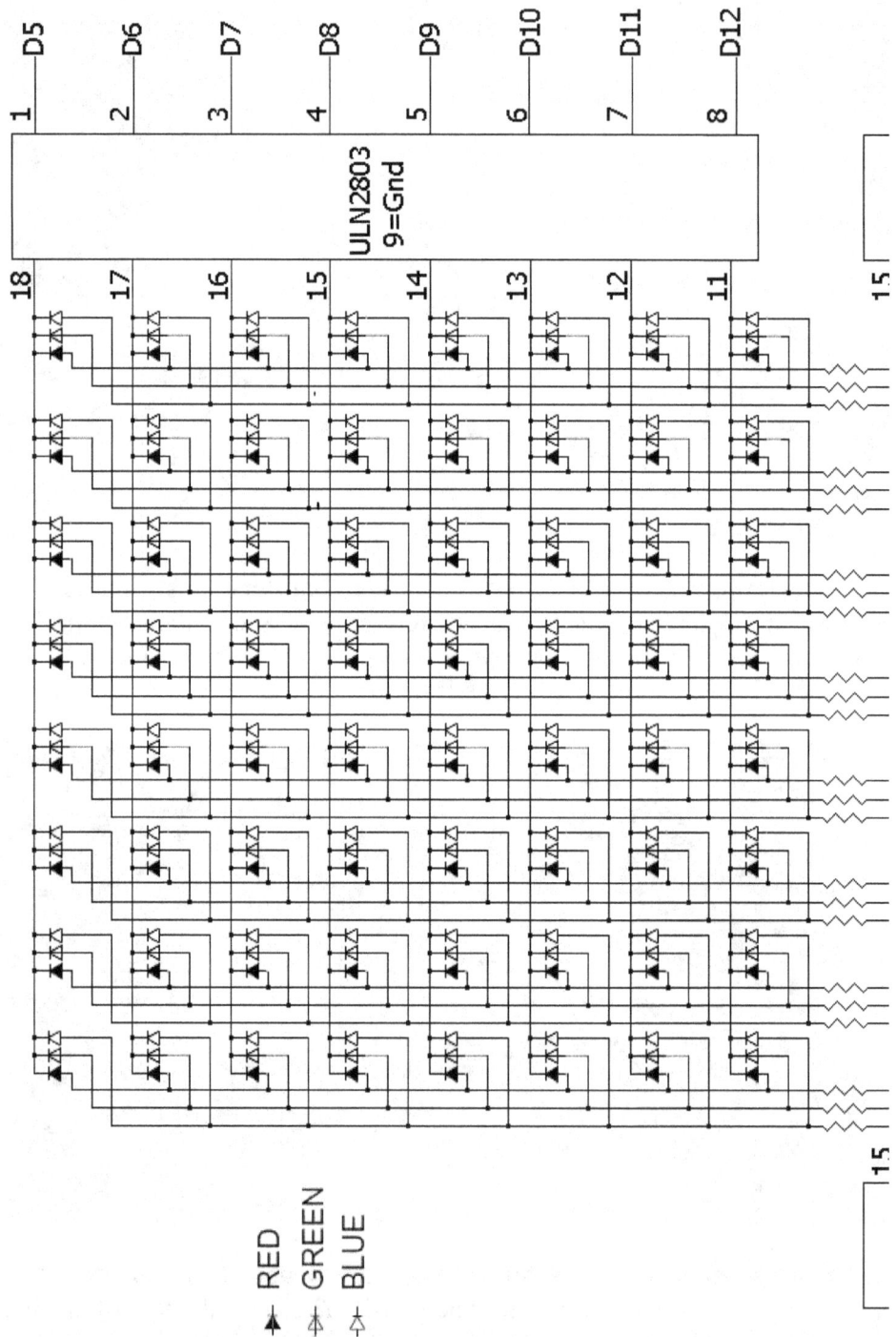

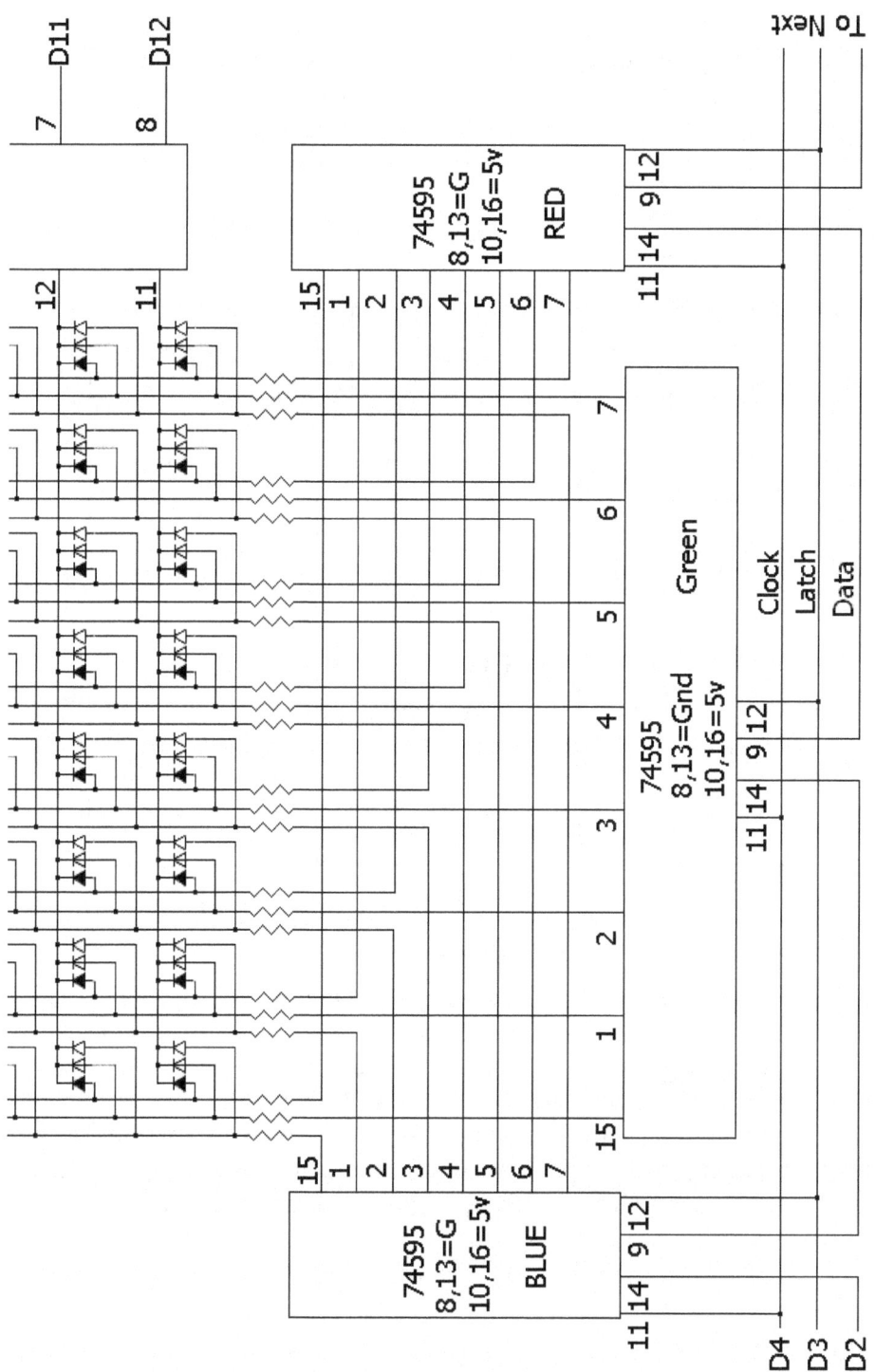

The schematic shows the red, green and blue shift registers wired in sequence. Initially I thought that there would be a red "in" a green "in" and blue "in" coming from the Arduino. The advantage would be that there would only be 64 "shifts" and it would be easy to separately control the colors. That design would look something like this drawing.

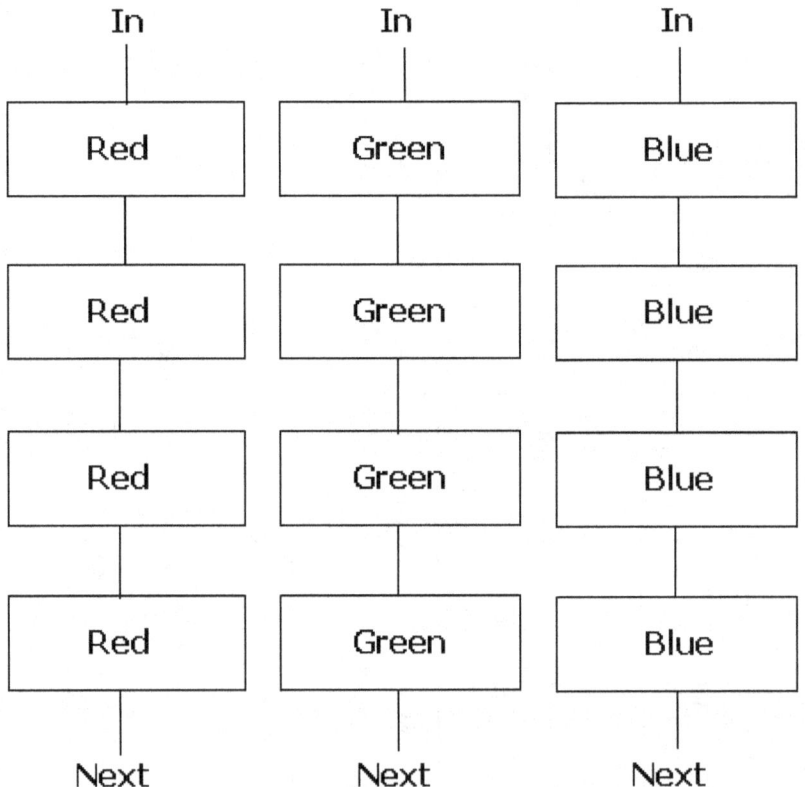

However the arrangement that was actually used put the red, green and blue registers in sequence. This arrangement is what was shown in the schematic diagram. This arrangement was chosen because some popular designs on the Internet used this setup. The red being last may look a little funny at first. However the first thing shifted in will end up at the far end of the shift registers. So then the blue being on the left and the red being on the right actually makes sense. This next diagram shows that arrangement.

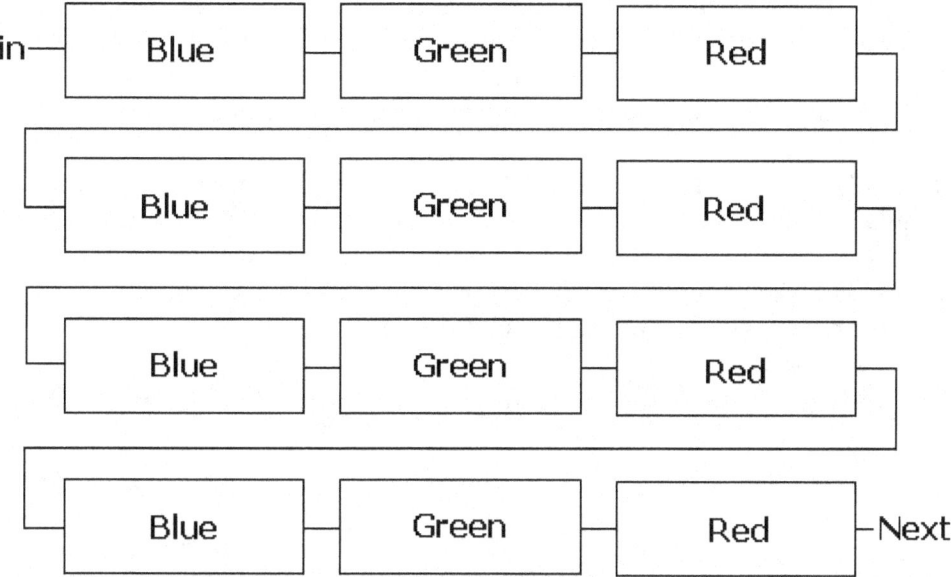

However I have found one design that put all of the blue shift registers together, then the green shift registers, and then the red shift registers are last. This arrangement looks something like this next diagram.

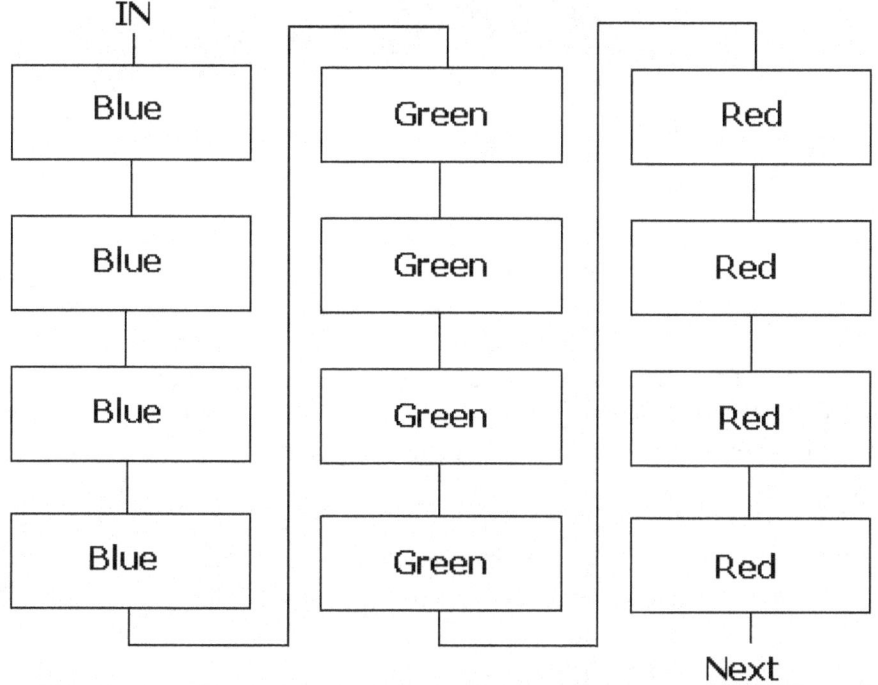

Another thing to consider in the arrangement of the shift registers is that most designs use a shift register to select the level as well. Usually this shift register is the first one in the chain so it is the last register to be loaded in the software. The advantage of this arrangement is that you do not need to turn the levels off to update the other shift registers. The "latch" command updates the level as well as all of the other shift registers at the same time.

The 8x8x8 color LED cube can be built on two 5x7 inch prototyping circuit boards. I decided to drill one extra hole for one end of each 8x8 array, but they could have been bent in slightly to fit as well. You will need two circuit boards as each one holds 8x8x4. Here is what one of the assemblies' looks like when it is built on a permanent circuit board.

On the bottom side of the circuit board the first thing you need to do is to run the power and ground busses. I use a 20 or 22 gage copper wire for the power busses. They need to be kept short and direct so they need to be run fist before all of the other wires are run. Next run any power or ground jumper wires such as those going to pins 10 and 13. Then you should run the shorter wires for instance each shift register can have two short wires to LED's.

The next picture shows the power and ground busses that are wired up to the IC sockets first.

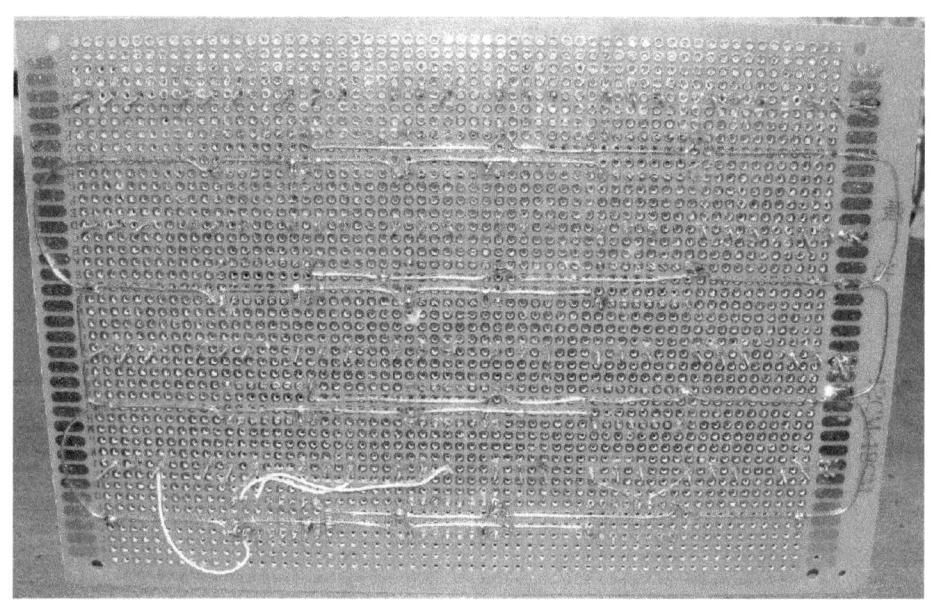

This next picture shows that pins 9 through 16 were wired up next. Pins 11 and 12 are a little tricky as they need to have two wires soldered to them. A trick in using 30 gauge wire is to only strip about a sixteenth of an inch of insulation. Then add solder to the IC socket pin and then poke the wire into that solder. If you strip too much insulation it will end up sticking out somewhere and will likely cause a short.

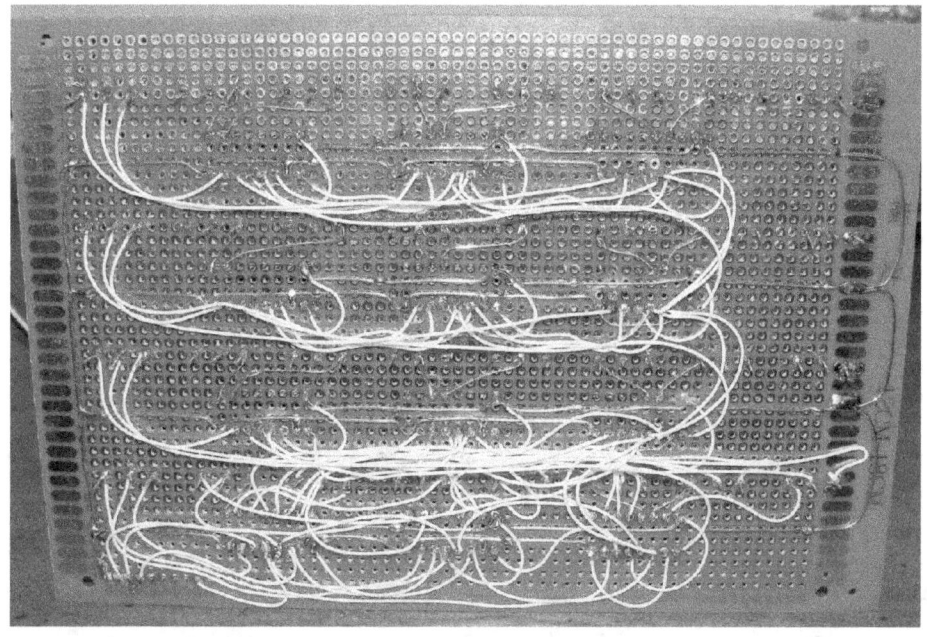

After a couple of weeks the circuit board is finally completed.

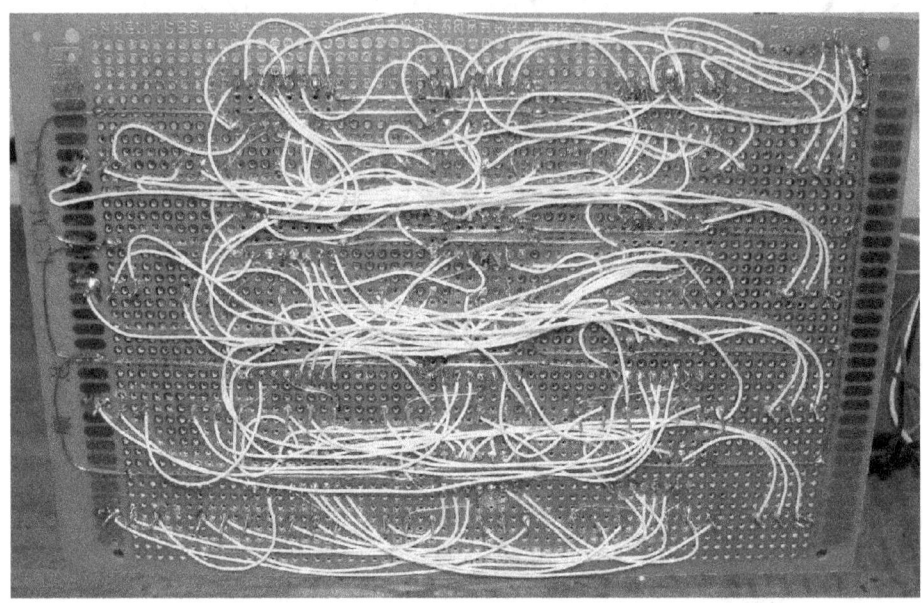

This first program listed here was used out test out the display to make sure everything worked. It turns all the layers red, then green, then blue, and then white. Turning all of the LED's white will require a one amp at 7 to 10 volt AC adapter to power the Arduino. If there is not enough power the Arduino will reset or the voltage will drop to about 3.5 volts and the LED's will be red instead of white.

```
// 8x8x8 RGB Cube testing all LED's
// These Pins Connect to 74595's
int data = 2;
int latch = 3;
int clock = 4;

// These Pins Connect to ULN2803
int Level1 = 5;
int Level2 = 6;
int Level3 = 7;
int Level4 = 8;
int Level5 = 9;
int Level6 = 10;
int Level7 = 11;
int Level8 = 12;
```

```
int rotate = 0;
int cycle = 0;

// set up output pins
void setup() {
  pinMode(data, OUTPUT);
  pinMode(clock, OUTPUT);
  pinMode(latch, OUTPUT);
  pinMode(Level1, OUTPUT);
  pinMode(Level2, OUTPUT);
  pinMode(Level3, OUTPUT);
  pinMode(Level4, OUTPUT);
  pinMode(Level5, OUTPUT);
  pinMode(Level6, OUTPUT);
  pinMode(Level7, OUTPUT);
  pinMode(Level8, OUTPUT);
}

void loop() {
  for (int level=0; level <8; level++){
    for (int shift=0; shift <96; shift++){
      digitalWrite(data, LOW);
      if (rotate == 0){
        if (shift >= 0 and shift < 8){digitalWrite(data, HIGH);}
        if (shift > 23 and shift < 32){digitalWrite(data, HIGH);}
        if (shift > 47 and shift < 56){digitalWrite(data, HIGH);}
        if (shift > 71 and shift < 80){digitalWrite(data, HIGH);}
      }
      if (rotate == 1){
        if (shift > 7 and shift < 16){digitalWrite(data, HIGH);}
        if (shift > 31 and shift < 40){digitalWrite(data, HIGH);}
        if (shift > 55 and shift < 64){digitalWrite(data, HIGH);}
        if (shift > 79 and shift < 88){digitalWrite(data, HIGH);}
      }
      if (rotate == 2){
        if (shift > 15 and shift < 24){digitalWrite(data, HIGH);}
        if (shift > 39 and shift < 48){digitalWrite(data, HIGH);}
        if (shift > 63 and shift < 72){digitalWrite(data, HIGH);}
        if (shift > 87 and shift < 96){digitalWrite(data, HIGH);}
      }
```

```
if (rotate == 3){
  digitalWrite(data, HIGH);
  }
  // Clocks in the new data
  digitalWrite(clock, LOW);
  digitalWrite(clock, HIGH);
}
// Turn the levels off
digitalWrite(Level1, LOW);
digitalWrite(Level2, LOW);
digitalWrite(Level3, LOW);
digitalWrite(Level4, LOW);
digitalWrite(Level5, LOW);
digitalWrite(Level6, LOW);
digitalWrite(Level7, LOW);
digitalWrite(Level8, LOW);

//Latches in the new data
digitalWrite(latch, LOW);
digitalWrite(latch, HIGH);
// Select the new level to turn on
if (level==0)digitalWrite(Level1, HIGH);
if (level==1)digitalWrite(Level2, HIGH);
if (level==2)digitalWrite(Level3, HIGH);
if (level==3)digitalWrite(Level4, HIGH);
if (level==4)digitalWrite(Level5, HIGH);
if (level==5)digitalWrite(Level6, HIGH);
if (level==6)digitalWrite(Level7, HIGH);
if (level==7)digitalWrite(Level8, HIGH);
  delay(1);
}
// Rate of movement - once per 50 cycles
cycle=cycle+1;
if (cycle == 50) {
  rotate = rotate+1;
cycle = 0;
}
if (rotate > 4) rotate=0;
}
```

This next program is the random number program that flashes random LED's and random colors. It is really bright!

```
// 8x8x8 RGB Cube with Random sparkles
// These Pins Connect to 74595's
int data = 2;
int latch = 3;
int clock = 4;

// These Pins Connect to ULN2803
int Level1 = 5;
int Level2 = 6;
int Level3 = 7;
int Level4 = 8;
int Level5 = 9;
int Level6 = 10;
int Level7 = 11;
int Level8 = 12;

int rotate = 0;
int cycle = 0;

// set up output pins
void setup() {
  pinMode(data, OUTPUT);
  pinMode(clock, OUTPUT);
  pinMode(latch, OUTPUT);
  pinMode(Level1, OUTPUT);
  pinMode(Level2, OUTPUT);
  pinMode(Level3, OUTPUT);
  pinMode(Level4, OUTPUT);
  pinMode(Level5, OUTPUT);
  pinMode(Level6, OUTPUT);
  pinMode(Level7, OUTPUT);
  pinMode(Level8, OUTPUT);
}

void loop() {
  for (int level=0; level <8; level++){
    int rdata1 = random(96);
    int rdata2 = random(96);
```

```
    int rdata3 = random(96);
    int rdata4 = random(96);
    for (int shift=0; shift <96; shift++){
      digitalWrite(data, LOW);
      if (rdata1==shift or rdata2==shift or rdata3==shift or rdata4==shift){
        digitalWrite(data, HIGH);
      }
      // Clocks in the new data
      digitalWrite(clock, LOW);
      digitalWrite(clock, HIGH);
    }
    // Turn the levels off
    digitalWrite(Level1, LOW);
    digitalWrite(Level2, LOW);
    digitalWrite(Level3, LOW);
    digitalWrite(Level4, LOW);
    digitalWrite(Level5, LOW);
    digitalWrite(Level6, LOW);
    digitalWrite(Level7, LOW);
    digitalWrite(Level8, LOW);

    //Latches in the new data
    digitalWrite(latch, LOW);
    digitalWrite(latch, HIGH);
    // Select the new level to turn on
    if (level==0)digitalWrite(Level1, HIGH);
    if (level==1)digitalWrite(Level2, HIGH);
    if (level==2)digitalWrite(Level3, HIGH);
    if (level==3)digitalWrite(Level4, HIGH);
    if (level==4)digitalWrite(Level5, HIGH);
    if (level==5)digitalWrite(Level6, HIGH);
    if (level==6)digitalWrite(Level7, HIGH);
    if (level==7)digitalWrite(Level8, HIGH);
    delay(5);
  }
}
```

This next program is the falling water program. This time it is also in random colors.

```
// 8x8x8 RGB Cube with falling rain
```

```
// These Pins Connect to 74595's
int data = 2;
int latch = 3;
int clock = 4;

// These Pins Connect to ULN2803
int Level1 = 5;
int Level2 = 6;
int Level3 = 7;
int Level4 = 8;
int Level5 = 9;
int Level6 = 10;
int Level7 = 11;
int Level8 = 12;

int rotate = 0;
int cycle = 0;

// stock up on random numbers
int rdata1 = random(96);
int rdata2 = random(96);
int rdata3 = random(96);
int rdata4 = random(96);
int rdata5 = random(96);
int rdata6 = random(96);
int rdata7 = random(96);
int rdata8 = random(96);
// set up output pins

void setup() {
  pinMode(data, OUTPUT);
  pinMode(clock, OUTPUT);
  pinMode(latch, OUTPUT);
  pinMode(Level1, OUTPUT);
  pinMode(Level2, OUTPUT);
  pinMode(Level3, OUTPUT);
  pinMode(Level4, OUTPUT);
  pinMode(Level5, OUTPUT);
  pinMode(Level6, OUTPUT);
  pinMode(Level7, OUTPUT);
  pinMode(Level8, OUTPUT);
```

```
}

void loop() {
  // stock up on random numbers.
  for (int level=0; level <8; level++){
    for (int shift=0; shift <96; shift++){
      digitalWrite(data, LOW);
      // select the random number for the correct level
      if (level==0 and rdata1==shift) digitalWrite(data, HIGH);
      if (level==1 and rdata2==shift) digitalWrite(data, HIGH);
      if (level==2 and rdata3==shift) digitalWrite(data, HIGH);
      if (level==3 and rdata4==shift) digitalWrite(data, HIGH);
      if (level==4 and rdata5==shift) digitalWrite(data, HIGH);
      if (level==5 and rdata6==shift) digitalWrite(data, HIGH);
      if (level==6 and rdata7==shift) digitalWrite(data, HIGH);
      if (level==7 and rdata8==shift) digitalWrite(data, HIGH);
      // Clock in the new data
      digitalWrite(clock, LOW);
      digitalWrite(clock, HIGH);
    }
    // Turn the levels off
    digitalWrite(Level1, LOW);
    digitalWrite(Level2, LOW);
    digitalWrite(Level3, LOW);
    digitalWrite(Level4, LOW);
    digitalWrite(Level5, LOW);
    digitalWrite(Level6, LOW);
    digitalWrite(Level7, LOW);
    digitalWrite(Level8, LOW);

    //Latches in the new data
    digitalWrite(latch, LOW);
    digitalWrite(latch, HIGH);
    // Select the new level to turn on
    if (level==0)digitalWrite(Level1, HIGH);
    if (level==1)digitalWrite(Level2, HIGH);
    if (level==2)digitalWrite(Level3, HIGH);
    if (level==3)digitalWrite(Level4, HIGH);
    if (level==4)digitalWrite(Level5, HIGH);
    if (level==5)digitalWrite(Level6, HIGH);
    if (level==6)digitalWrite(Level7, HIGH);
```

```
    if (level==7)digitalWrite(Level8, HIGH);
    delay(.5);
  }
// Rate of movement - one per 8 cycles
rotate = rotate+1;
if (rotate > 7) {
  rotate=0;
  // shift the random numbers down
  rdata8 = rdata7;
  rdata7 = rdata6;
  rdata6 = rdata5;
  rdata5 = rdata4;
  rdata4 = rdata3;
  rdata3 = rdata2;
  rdata2 = rdata1;
  rdata1 = random(96);
  }
}
```

To make this RGB cube compatible with software that was written for other RGB cubes we will need to make some changes. First we need to add a 74595 to select the layer. Then we need to change the data, clock, and latch pins. Up next is an updated schematic showing all of the necessary changes.

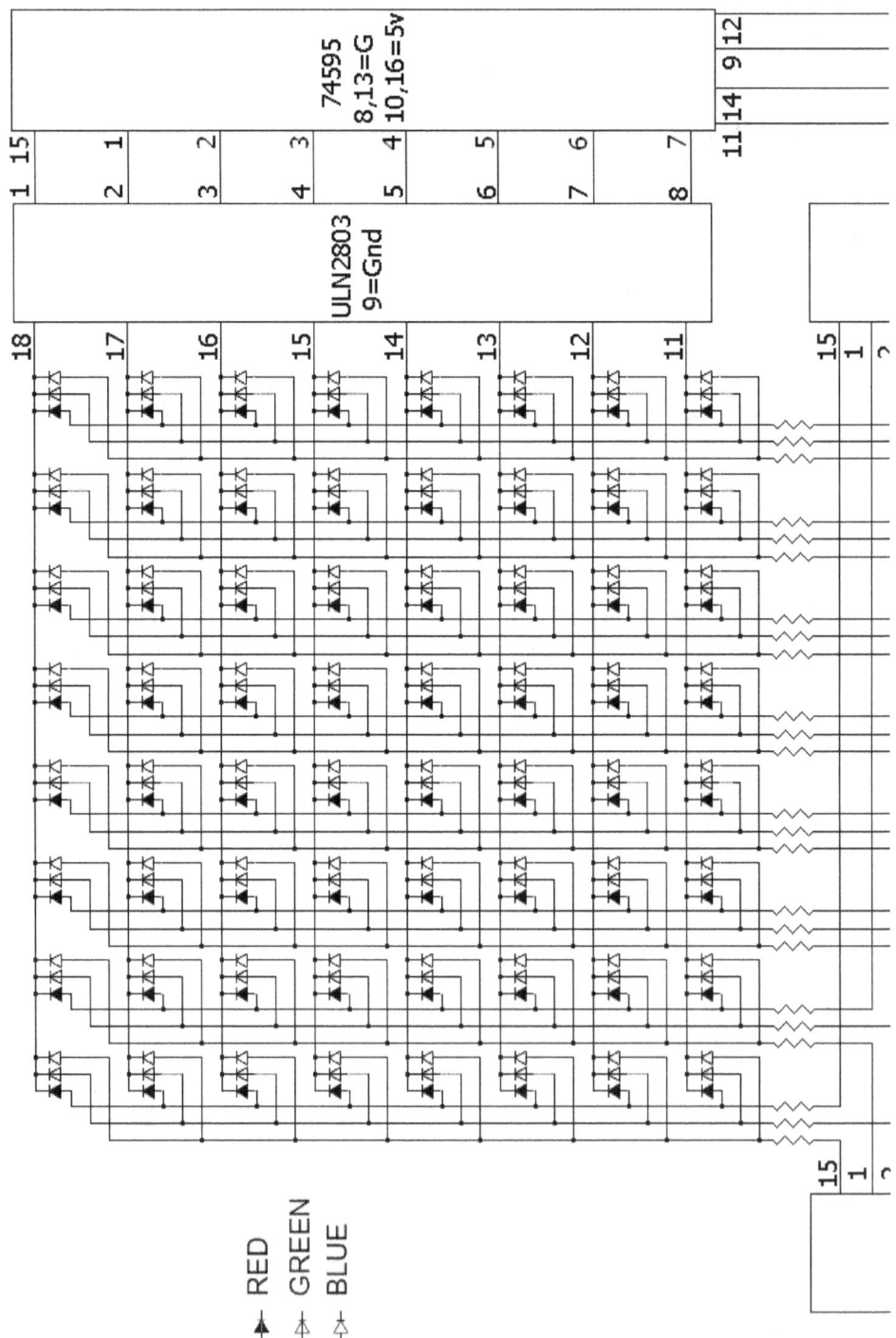

74595
8,13=G
10,16=5v

ULN2803
9=Gnd

RED
GREEN
BLUE

90

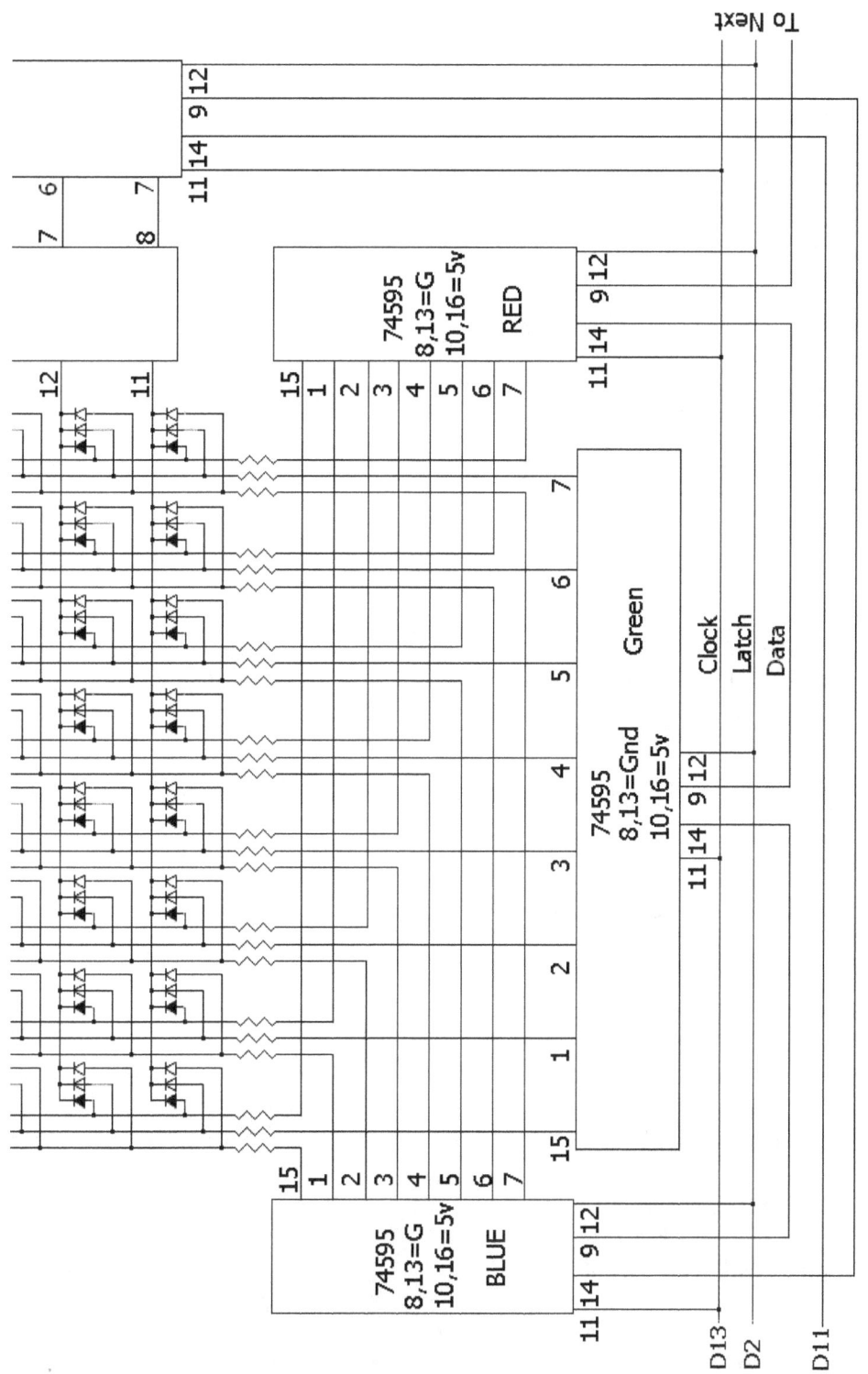

Here is some code to test the rearranged RGB cube. Using a shift register to select the level makes the program a lot shorter.

```
// 8x8x8 RGB Cube with Random sparkles
// This version uses a 74595 to select the level
// These Pins Connect to 74595's
// They are reassigned for compatibility.
int data = 11;  // 595 pin 14
int latch = 2;   // 595 pin 12
int clock = 13; // 595 pin 11
// set up output pins
void setup() {
  pinMode(data, OUTPUT);
  pinMode(clock, OUTPUT);
  pinMode(latch, OUTPUT);
}
void loop() {
  for (int level=0; level <8; level++){
    int rdata1 = random(96);
    int rdata2 = random(96);
    for (int shift=0; shift <96; shift++){
      digitalWrite(data, LOW);
      if (rdata1==shift or rdata2==shift){
        digitalWrite(data, HIGH);
      }
      // Clocks in the new data
      digitalWrite(clock, LOW);
      digitalWrite(clock, HIGH);
    }
    for (int lshift=0; lshift <8; lshift++){
      digitalWrite(data, LOW);
      if (lshift == level) digitalWrite(data, HIGH);
      // Clocks in the level data
      digitalWrite(clock, LOW);
      digitalWrite(clock, HIGH);
    }
    //Latches the new data
    digitalWrite(latch, LOW);
    digitalWrite(latch, HIGH);
    delay(5);
} }
```

Chapter 10

Parts Sources

Most of the parts are found on eBay.

Here are some 5 inch by 7 inch board for using instead of the breadboards for making a permanent version of any of the LED cubes.

1 item sold by luckycardinal

THREE 5x7" (12x18cm) Prototyping PCB Printed Circuit Board Prototype Breadboard
(120946639411)

Estimated delivery **Mon, May 18**
Tracking number: 9400109699937590183714

Here is the 8x8x8 LED kit. The LED's from this kit were used to make the 8x8x8 LED cube as well as for the kit itself.

1 item sold by princessdress08

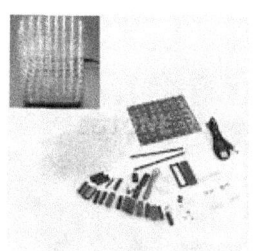

3D LightSquared 8x8x8 LED Cube White LED blue Ray DIY New Kit
(141503417127)

Delivered on **Mon, Mar 23**
Tracking number: LK400355530CN

Here are the breadboards that can be taken apart to make the 8x8x8 cube.

MB-102 Solderless Breadboard Protoboard 830 Tie Points 2
buses Test Circuit F5
(291369269558)

ⓘ Estimated delivery **Tue, Apr 28 - Tue, May 12**

Here are the 5mm Common Anode RGB LED's. These were used to make
the 4x4x4 Color LED cube. I ordered these by accident but they might make
a brighter color cube.

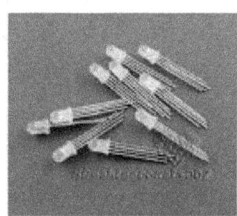

10~1000Pcs Lots 5mm 4Pin RGB Common Anode Cathode
LED Lamp Emitting Light Diode
(261563947961)

Type: Diffused
Packing quantity: 100pcs
Colors: RGB Common Cathode
Size: 5MM

Here are the 5mm common anode RGB LED's that I used for the 8x8x8 RGB
cube.

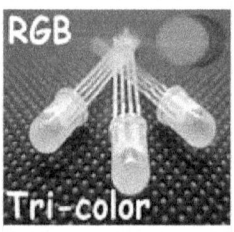

50 pcs 5mm 4pin Diffused RGB Tri-Color Common Cathode
Red Green Blue LED
(400498119840)

Quantity: 2

ⓘ Estimated delivery **Wed, May 06 - Wed, May 20**
   Tracking number: **LK453977006CN**

Bibliography

Some 4x4x4 direct drive cube designs:

http://www.instructables.com/id/4x4x4-LED-Cube-Arduino-Uno/
http://www.instructables.com/id/4x4x4-LED-cube-based-Arduino-and-Flower-protoboard/

A 4x4x4 direct drive cube code generator:

http://www.instructables.com/id/LED-CUBE-CODE-4x4x4-Arduino/

An 8x8x8 LED Cube kit

https://icstation13.wordpress.com/2014/09/29/3d-lightsquared-8x8x8-led-cube-soldering-steps-2/

An 8x8x8 LED Cube with Arduino and latches

http://www.pyroelectro.com/projects/8x8x8_led_cube/

A 4x4x4 Color LED Cube

http://www.instructables.com/id/4x4x4-RGB-LED-Cube/

An 8x8x8 Color LED Cube

http://www.instructables.com/id/RGB-8x8x8-LED-Cube-1/

www.ingramcontent.com/pod-product-compliance
Lightning Source LLC
Chambersburg PA
CBHW080829180526
45168CB00006B/2615